D0207863

McGill-Queen's Rural, Wildland, and Resource Studies Series
Series editors: Colin A.M. Duncan, James Murton, and R.W. Sandwell

The Rural, Wildland, and Resource Studies Series includes monographs, thematically unified edited collections, and rare out-of-print classics. It is inspired by Canadian Papers in Rural History, Donald H. Akenson's influential occasional papers series, and seeks to catalyze reconsideration of communities and places lying beyond city limits, outside centres of urban political and cultural power, and located at past and present sites of resource procurement and environmental change. Scholarly and popular interest in the environment, climate change, food, and a seemingly deepening divide between city and country, is drawing non-urban places back into the mainstream. The series seeks to present the best environmentally contextualized research on topics such as agriculture, cottage living, fishing, the gathering of wild foods, mining, power generation, and rural commerce, within and beyond Canada's borders.

Nature, Place, and Story

Rethinking Historic Sites in Canada

CLAIRE ELIZABETH CAMPBELL

McGill-Queen's University Press

Montreal & Kingston • London • Chicago

© McGill-Queen's University Press 2017

ISBN 978-0-7735-5125-1 (cloth)
ISBN 978-0-7735-5177-0 (ePDF)
ISBN 978-0-7735-5178-7 (ePUB)

Legal deposit third quarter 2017
Bibliothèque nationale du Québec

Printed in Canada on acid-free paper that is 100% ancient forest free (100% post-consumer recycled), processed chlorine free

This book has been published with the help of a grant from the Canadian Federation for the Humanities and Social Sciences, through the Awards to Scholarly Publications Program, using funds provided by the Social Sciences and Humanities Research Council of Canada.

McGill-Queen's University Press acknowledges the support of the Canada Council for the Arts for our publishing program. We also acknowledge the financial support of the Government of Canada through the Canada Book Fund for our publishing activities.

Library and Archives Canada Cataloguing in Publication

Campbell, Claire Elizabeth, 1974–, author
Nature, place, and story : rethinking historic sites in Canada / Claire Elizabeth Campbell.

(McGill-Queen's rural, wildland, and resource studies series ; 8)
Includes bibliographical references and index.
Issued in print and electronic formats.
ISBN 978-0-7735-5125-1 (cloth).
ISBN 978-0-7735-5177-0 (ePDF)
ISBN 978-0-7735-5178-7 (ePUB)

1. Historic sites – Canada. 2. Public history – Canada. 3. Canada – Environmental conditions – History. I. Title. II. Series: McGill-Queen's rural, wildland, and resource studies series ; 8

FC215.C33 2017   971   C2017-902125-7
                                     C2017-902126-5

Contents

Acknowledgments

This book has been a long time in the making. The first acknowledgment must be to a certain VW Westfalia van, my parents, and the years we spent driving around North America exploring parks, historic sites, and landmarks. But as something more than family slide shows, this project began with a postdoctoral fellowship at the University of Alberta, with the warm support of the Department of History and Classics and the special generosity of Gerhard Ens. A year spent working with Alberta's Historic Resources Management Branch in Edmonton supplied numerous conversations with Larry Pearson and others about the meaning of historical significance and historic places. A start-up grant from Dalhousie University took me to Winnipeg, where I fell in love with the Forks during a July heat wave, while teaching in Dal's College of Sustainability introduced me to the marshlands at Grand Pré and the place for historians in questions of environmental education and policy. A sabbatical fellowship at the McGill Institute for the Study of Canada and a seminar in environmental history there jump-started the writing process. Ironically, most of this book was written from away, as they say in the Maritimes. I'm grateful to Bucknell University for financial support since my arrival in 2013, and to my colleagues in the Department of History who have prompted me to see nationhood, politics, and academic engagement differently.

Over the years, I've benefitted from feedback from colleagues from different fields at any number of gatherings, from Turku to Vancouver. Above all, I am grateful to Alan MacEachern and others in the Network in Canadian History & Environment (NICHE) for all manner of support and encouragement. NICHE – formerly a SSHRC-funded

Strategic Cluster – transformed the field of environmental history in Canada: it created unprecedented opportunities for aspiring historians, especially for us to engage as scholars and citizens with the paramount issue of our era. In addition, I want to acknowledge the historians who work for Parks Canada; they have a much harder job than I do, and I admire enormously the way that they model public service and scholarship in often challenging circumstances.

My editors at McGill-Queen's University Press have been wonderful: warm, supportive, and inspiring in their commitment to academic publishing. Jacqueline Mason was a voice of wisdom and encouragement (and patience, as I moved to another country and had a baby) throughout the writing process, while Ryan Van Huijstee offered guidance and made the book a reality. I'm also grateful to the anonymous reviewers who devoted time and energy to offering constructive suggestions on the manuscript.

Lastly, to Anthony and Campbell Jesse. I've left bits of myself across Canada in writing this book, but you are home.

Permissions

A version of Chapter 2 was published as "Idyll and Industry: Rethinking the Environmental History of Grand Pré, Nova Scotia," in a special issue of the *London Journal of Canadian Studies* 31 (UCL Press, Autumn 2016).

Nature, Place, and Story

Historic Sites in Canada and the Place of Environmental History

In looking at nature, I was tracking down history.
– Babette Mangolte, *The Sky on Location* (1982)

Why This Book? Why These Places?

My father taught high-school history, geography, and civics for forty years, which meant – and my apologies to all teachers for what comes next – that he got the summers "off." But while you can take the teacher out of the classroom for two months, you can't take the instinct to teach out of the teacher. When I was eight years old, he bought a Volkswagen Westfalia camper, and over the next several summers drove our family all over North America, from Cape Breton to Florida, from the north shore of Lake Superior to Vancouver Island to South Dakota, visiting historic sites, national parks, and other landmarks and attractions. We ate pea soup in the fog at the Fortress of Louisbourg. We learned that Pancake Bay was named by voyageurs who mixed the remnants of flour left in their packs with lake water for meals. We saw Scotts Bluff against the Nebraska sky in our camper just as those in wagon trains sought landmarks westward on the Oregon Trail. So "history" was three-dimensional for me right from the start, associated with tangible sensations and different environments. I was converted early and completely to the belief that history is affected by where it happens, and that learning history *in situ*, or "in place," can be extraordinarily powerful.

Historic sites, though, are more than fun places for family vacations, and more than convenient destinations for school fieldtrips. The story of historic sites in Canada is that of a relatively young country trying to create a coherent national story for itself, its varied and often divided population, and the world. Historic sites form a place-library of physical texts that record changing ideas about our history, and how we have told our story in the past. We can see that library expand from battlefields of the wars between French and British empires that spoke to British conquest and a heritage of "two nations," to sites that commemorate industrial plants and immigrant communities. So most of the scholarship about historic sites has focused on what they tell us about the politics of public history, commemoration, and the construction and buttressing of a national identity. But historic sites record another story: our history of occupying, transforming, and adapting to new environments and landscapes *a mari usque ad mare*, from sea to sea. Individual sites present local discoveries of specific places, while the system as a whole – our "national heritage" – celebrates a collection of ecological possessions of enormous range and diversity. Given the crucial role of environmental issues in the present, we need more on their past.

This book visits five large and prominent historic landscapes across Canada to ask how we might begin to do that. All are national historic sites; some are World Heritage Sites. All include substantial and diverse environmental features, from oceanic coastline to boreal forest to rolling grasslands. And all can be reimagined as stories of environmental change as well as the making of Canada. Each chapter here outlines a site's environmental character and history, and its current depiction of the Canadian national story. But together they ask how we might use these places to teach the history of Canada's peoples *within* Canada's environments, and to connect these histories with discussions that Canadians are having today about the direction in which we are headed *as* a people: a braided narrative of natural history, culture, and political choices.[1]

- L'Anse aux Meadows, "the gateway to the New World," is the only known site of Norse settlement in North America, and a place to explore the long-term effects of two of the most critical issues in marine environments: climate change and species exhaustion.

- Grand Pré's fertile farmlands drained from the Bay of Fundy's saltwater marshes by French-speaking Acadians in the seventeenth and eighteenth centuries also represents the mechanisms and costs of industrial agriculture in the Annapolis Valley.
- At the northwest tip of Lake Superior, Fort William was one of the most important fur trade posts in Canadian history. Its reconstruction in the 1970s says much about public and political expectations of wilderness and authenticity.
- In contrast, the Forks of the Red River in Winnipeg was rehabilitated as an urban park, one that tries to represent multiple periods of historical occupation, but serves primarily as green space to complement a gentrified post-industrial downtown market economy.
- Set amid Alberta's unbearably beautiful foothills, the Bar U Ranch invokes a romantic image of frontier, even though ranches were and are part of a political/corporate axis of national expansion and a transcontinental network of food production. And there has been oil and gas production in the area for almost as long as there have been ranches, signalling our commitment to an extractive economy and faith in a resource frontier.

Together, these represent five of the largest public history projects undertaken in Canada since the 1950s, and are among the most iconic of Canada's historic places, symbols of both a national history and the different regions to which they belong. Moving from east to west, we move through two chronologies: the history of Canada as a settler nation, from the arrival of the Norse a thousand years ago to corporate agriculture in the mid-twentieth century; and different ways of thinking about the country's past, from a heroic narrative of discovery to a more complicated, somewhat less-sure account of regional and cultural diversity. Each site has been preserved, rediscovered, altered, relocated, or reconstructed in a different way. This lets us ask comparative questions about public history as it affects and involves the natural environment. Where and why do historic sites reference their surroundings? What kinds of stories, places, and artifacts are most adaptable to change – ecological change as well as changing interpretations? What is the effect of selling history as place? Is it possible to teach about an environmental past through a human lens? And can this past make an intervention in the present? These five sites were chosen because they involve substantial tracts of space, but also because they represent transnational scope and ecological variety, a range of environmental issues, and different responses to environmental as well as historical resources. They are caught up in the most prominent environmental questions in public debate today: climate change, agricultural sustainability, wilderness protection, urban reclamation, and oil and gas extraction.

Several years ago I stumbled across a poem titled "A Message to Winnipeg," by James Reaney (1926–2008), a Governor General Award–winning writer from Stratford, Ontario. It begins,

> Winnipeg, what once were you? You were,
> Your hair was grass by the river ten feet tall,
> Your arms were burr oaks and ash leaf maples,
> Your backbone was a crooked silver muddy river,
> Your thoughts were ravens in flocks, your bones were snow,
> Your legs were trails and your blood was a people
> Who did what the stars did and the sun.[2]

Just as Reaney mentally erased the built landscape of mid-century Winnipeg to ask what and who existed there before, we need to reimagine the older landscapes that inspired, influenced, and framed human decisions in the past, to understand not only how our history took shape but the origins of the environmental issues which we now face.

The Great Divide: History and Environment in Protected Places

In 2009, historical geographers Graeme Wynn and Matthew Evenden wrote, "there have been few efforts to situate canonical events and problems in Canadian history within an environmental context. What

0.1 H.L. Hime, *Tents on the Prairie, West of the Settlement, Red River, Manitoba*, 1858. (Courtesy of the McCord Museum N-0000.68.1)

do environmental historians have to say about the building of the rail-road, the growth of the welfare state or Quebec nationalism?"[3] Their choice of examples reflects the recent, and westward, emphasis of most writing about Canadian history of late, but the same could be asked of the explorations of the North Atlantic by the Norse in the tenth century, or the deportation – *le grand dérangement* – of the Acadians in the mid-eighteenth. Much environmental history to date has been in-tensely local, but a series of recent and ambitious books have attempted to synthesize a national environmental history.[4] Because for all our dif-ferences, we do share a federal system of governance, not least in heritage and the environment, and a canon of images taken *a mari usque ad mare*. What we need is a conversation between and across regions, to listen to other experiences from across the country. A syllabus for "A History of Canada" might look very different if it focused on the de-velopment of resource industries, exploration and settlement against coastal and climatic change, or nature in arts and culture.

Despite the remarkable growth of environmental history in North America in the past few decades, the narratives of national history and environmental change frequently exist as two solitudes between humanists and scientists. This separation is embodied in our protected places, divided as they are between nature and history, between national parks and historic sites, and between natural and cultural World Heritage Sites. This dichotomy was neatly symbolized in 1978 with Canada's first two World Heritage Sites: L'Anse aux Meadows on the one hand, Nahanni National Park on the other. Since the 1970s, Parks Canada has acknowledged the human imprint in national parks by characterizing them as cultural landscapes rather than wilderness. But there has been no answering movement on the other side of the spectrum, where historic sites remain profoundly anthropocentric. According to the *National Historic Sites System Plan*, sites must

- illustrate an exceptional creative achievement in concept and design, technology or planning, or a significant stage in the development of Canada;
- illustrate or symbolize, in whole or in part, a cultural tradition, a way of life or ideas important to the development of Canada;
- be explicitly and meaningfully associated or identified with persons who are deemed to be of national historic significance;
- or be explicitly and meaningfully associated or identified with events that are deemed to be of national historic significance.

The environment is generally framed as backdrop to heroic human endeavour, or raw material for human innovation – what Parks Canada calls "human creativity."[5] The land is acted upon, not actor, in the story of building a nation.

But these sites were never the result *solely* of human ingenuity or effort. That effort was guided, sometimes encouraged, sometimes thwarted, by the environment in which people found themselves. And while this emphasis on human action lends itself to colourful and dramatic stories, it is a blinkered and highly problematic view of Canadian history if we are to explain the situation in which we find ourselves. As Thomas Symons, then outgoing-chair of the Historic Sites and Monuments Board of Canada (HSMBC), wrote in 1997,

On doit donc se demander pourquoi le patrimoine ne constitue pas un volet du movement de defense de l'environnement ... Beaucoup de nos lieux historiques et de nos commémorations ont semblé souligner l'exploitation de l'environnement, car nous célébrions le mauvais aspect de realisations matérielles. [We must therefore ask why heritage is not a part of the environmental movement ... Many of our historic sites and our commemorations seemed to emphasize the exploitation of the environment, as we celebrate the wrong side of material achievements.][6]

Sites that heroize imperial and industrial expansion without environmental context quietly sanction and perpetuate the same ethos in contemporary Canada. A recent advertising campaign by oil company Cenovus showed an unbroken expanse of boreal forest overlain with a chronology of Canadian technological achievements: the telephone, the transcontinental railway, the Canadarm. The message is that the land inspires our actions but is never affected or harmed by them – a message that is simply untrue. We need to stop divorcing the historical record from current economic activity, a rhetorical romance of wilderness from active resource development. Yet our system of protected places has merely encouraged this self-deception. This is precisely the paradoxical raison d'être we have always assigned to our national parks: "dedicated to the people of Canada for their benefit, education, and enjoyment [while] maintained and made use of so as to leave them unimpaired for future generations."[7]

World Heritage Sites, despite their deceptively neat categorizations of cultural and natural, emerged from a realization that heritage and environment are fundamentally entwined. In 1972, the UNESCO Convention Concerning the Protection of the World Cultural and Natural Heritage proposed a "system of collective protection of the cultural and natural heritage of outstanding universal value." This founding convention cited concern for the "deterioration" and "disappearance" of places "not only by the traditional causes of decay, but also by changing social and economic conditions which aggravate the situation with even more formidable phenomena of damage or destruction." While criteria like biodiversity and natural beauty are generally applied to "natural" sites, UNESCO recognizes the environmental context of the

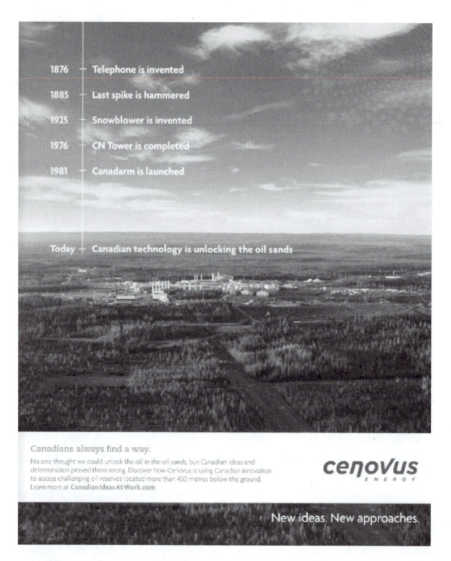

0.2 Cenovus ad, 2012. (Public domain)

"works of man" (in other words, "human creativity"), and how some
of those works are in fact deliberate environmental impacts. A site can
be designated to represent "as an outstanding example of a traditional human settlement, land-use, or sea-use which is representative
of a culture (or cultures), or human interaction with the environment

especially when it has become vulnerable under the impact of irreversible change."[8] Part of our heritage, in other words, is the human experience of environmental change, which may include causing a new state of vulnerability. Heritage designation can and should demand a frank discussion about sustainable and unsustainable practices in history. Of course, environmental sustainability was not a concern for Canada's historic sites system for much of its history. But this system has always responded to changes in the country's political, social, and intellectual landscape. It can do so again.

Building a (Story for a) Nation: Assembling Canada's Historic Sites

Scholarship on public history and heritage has focused on the politics of commemoration: how and what people decide to remember and celebrate; who is involved in or excluded from the decisions; how this changes over time; and how physical monuments become *lieux de memoire* or "structural supports" of public memory.[9] There has been much more written on the ideologies behind museum collections, on the one hand, and human history in national parks, on the other, than on historic sites. Indeed, the only full-length work on Canada's national historic sites remains C.J. Taylor's 1990 *Negotiating the Past: The Making of Canada's National Historic Parks and Sites*. And there have been only a few efforts to raise questions about how public history intersects with the environment.[10]

In fact, historic sites in Canada and the United States are built on a twinned tradition of natural and human architectures, of physical and designed monuments. Local history societies proliferated in the latter nineteenth century, but the first significant state commitment came with the commemoration in 1908 of the Plains of Abraham – the quintessential historic landscape – as a national landmark. (This came one year after the American Antiquities Act granted executive authority to the president of the United States "to declare by public proclamation historic landmarks.") In 1919, Ottawa established an advisory board, the forerunner to the Historic Sites and Monuments Board, within the Dominion Parks Branch, reflecting the close relationship perceived to be between artifact and setting. For much of the twentieth century, the theme of land was crucial, yet relatively unexamined. The board favoured sites that present a coherent narrative of Canada's territorial expansion, as the geographical stage for a reconciliation of the two founding empires (Indigenous lands being considered inconsequential) and the coming of peace, order, and good government, with dashes of wilderness excitement and technological triumph to enliven the story. Bronze plaques on stone cairns were placed near the ruins of star-shaped forts to mark the struggle for a continent between France and Britain, and later efforts to defend loyalist British North America from our acquisitive neighbours to the south; on the site of (often long-vanished) fur trade posts from Hudson Bay to the Pacific; or places where North West Mounted Police sought to pacify the "rebel" Métis.[11] These were designed to explain the origins of Canada's borders as a transcontinental "dominion of the north": where the singular political aim of national expansion drew from, and held together, vastly different physical resources.

Meanwhile, James B. Harkin, the first commissioner of the Dominion Parks Branch, envisioned a new category of historic parks as a way of expanding the national parks system eastward from Banff and Jasper, and amplifying the Branch's profile in the federal bureaucracy. As early as 1917, the Branch acquired Fort Anne in Nova Scotia – an early eighteenth-century British fort built atop a seventeenth-century French settlement – to "provide a landscape evocative of a range of historic associations and a recreational area for local citizens and tourists." Historic parks were imagined as accessible and scenic green spaces enclosing evocative and hopefully educational ruins, which would act as "focal points and attractions to the suburban park." Such "focal points and attractions" lent themselves to further development, very appealing to a Parks Branch intent on generating tourism.[12] The 1930 National Parks Act stated that

> 11. The Governor in Council may set apart any land, the title to which is vested in His Majesty, as a National Historic Park to
> (i) commemorate an historic event of national importance, or
> (ii) preserve any historic landmark or any object of historic, prehistoric or scientific interest of national importance.

South of the border, the National Parks Service had a similar interest in acquiring battlefields and other historic landmarks, though national building surveys remained focused on architectural styles and occupants.[13] Inspired by the massive reconstruction of Colonial Williamsburg and the use of historic sites as public works projects during the Great Depression, Ottawa funded partial reconstructions at larger sites like the Halifax Citadel and Prince of Wales Fort in hopes of stimulating both the economy and patriotic sentiment. In both countries, a national history served to enhance federal authority, and vice versa.

But the historic site as we know it is really a phenomenon of the political, demographic, and commercial agendas of postwar North America.[14] By the 1950s, national historic sites were widely recognized as attractions for visitors from at home and abroad as well as symbolic statements of a distinctively Canadian (not-American) story. The Royal Commission on National Development in the Arts, Letters and Sciences, also known as the Massey Commission, found it "curious" that historic sites were the responsibility of the Parks Branch, but accepted that this arrangement was designed to appease "consumer interest." But it cautioned that "the principal object ... should be to instruct Canadians about their history through the emotional and imaginative appeal of associated objects. Factual information can be obtained in books; the function of the monument or marker is, we assume, to convey a sense of the reality of the past."[15] The Parks Branch, though, felt that "imaginative appeal" required some material assistance, and reconstructions accelerated through the 1960s, supported by massive funding for Centennial projects, a corresponding concern for solidifying a national history, and the growing popularity of living history programs with a (baby-)booming audience. At fur trade sites across the country, visitors could expect storehouses stocked with familiar artifacts such as bundles of furs and Hudson's Bay Company blankets. The Fortress of Louisbourg rose again on the coast of Cape Breton, the largest reconstruction ever undertaken in Canada, a triumph for – depending on your point of view – the field of Canadian archaeology, Parks' engineers, or federal employment programs. Nature was critical to the success of these sites, but in unquestioned ways, primarily as setting or ambiance. Acadians prospered quietly in rich and idyllic farmland; the explorer and *voyageur* tested his mettle in the boreal wilderness; the soldier on watch patrolled the ramparts overlooking the approach to Quebec or the Gulf of the St Lawrence.[16]

But Louisbourg was an anachronism even before it was completed. A national framework, geographic or historical, came under attack by historians who emphasized the "limited" identities of region, class, and gender instead of a singular national identity. To borrow Cole Harris's wonderful metaphor, the illusion of Canada as a single entity dissolved into an archipelago.[17] Ottawa retreated from funding historical mega-projects at the same time as historians retreated from a national narrative. Archaeological research, previously used to inform reconstructions, now became a showpiece in its own right, whether unearthing the tree stumps left from shelterbelts on prairie farms or marking the wreck of a sixteenth-century galleon in Labrador. Historic sites were more likely to be a collection of artifacts left *in situ* to convey "a sense of place" than reconstructions designed to transport the visitor back in time. This, of course, was both pedagogically and fiscally prudent – less heavy-handed in assigning a fixed story to a particular site, ostensibly more authentic, and much, much cheaper.

It also suited the tense political arena of federal–provincial relations. Historic sites remained with the National Parks Branch, separate from the growing constellation of national museums (in Ottawa), and thus theoretically, if not actually, closer to discussions about regional ecologies. Provinces, from Quebec to the rising "new west," were ever more assertive in arts, culture, heritage, parks, and, most importantly, natural resources. Unsurprisingly, a 1982 federal report on cultural resources – after years of exhausting constitutional wrangling – noted that "interests, aspirations and sensitivities differ widely from region to region. It is entirely reasonable that institutions in each region should develop collections and exhibitions which reflect the distinctive characteristics of that region."[18] Exercising restraint and acknowledging regional difference was, for a federal agency, prudent.

There was also a new concept pervading thinking about historic sites: the cultural landscape. A sudden flurry of park creation in the early 1970s, especially in the Far North, coincided with a new political presence by Canada's First Nations. This required the Parks Branch to reconsider the idea of national parks as uninhabited or pristine wilderness, and recognize the human occupants and ongoing cultural traditions in these places. (Indigenous lands remain the most common illustration of cultural landscape in Canada, because of their depth of historical association and because they are easily depicted as an alternative to the Enlightenment dichotomy of reason and nature.[19] But this itself is a dichotomy, and may slow us from seeing that settler culture, too, relies on the natural world for material and psychological sustenance.) Historic sites, especially living history programs, did not move toward environmental history and ecological content in the same way, but there were signs that the system was evolving toward landscape thinking. In the mid-1980s, the HSMBC recognized various settlement patterns as having national significance, such as the Métis river lot, the Mormon irrigation plat, and leasehold tenure on Prince Edward Island. While it was still easiest to plaque a grand Victorian home, new sites might be complex groupings of buildings, as with twentieth-century industrial plants, such as the Gulf of Georgia Cannery and the Gooderham and Worts Distillery; or archaeological sites, like Melanson Settlement or Port au Choix. The new designation of "national historic district" was another way to incorporate larger or more complex areas, like urban neighbourhoods or places with multiple periods of significance, such as the town of Lunenburg.

If the cultural landscape acknowledged a human imprint, it also managed to convey an environmental(ist) sensibility. In Canada and the United States, national parks and monuments had been a logical, and convenient, starting point for conservationists for over a century. Both countries used these categories of federal lands to expand federal authority through public works as well as experiment with targeted conservation projects for particular species. By the 1970s, however, there was growing support for environmentalist organizations, environmental legislation, and a more comprehensive approach to wildlands protection.[20] Trailing slightly behind the National Park Service, eventually Parks Canada reworked its operating guidelines to acknowledge that "national historic sites reflect the enormous impact that the biophysical environment – our natural heritage – has had on

the landscape of Canadian history." Site interpretation was now to discuss historic character and significance, but also "relevant links between historical activities and the natural environment." This shift was
likely influenced by heightened anxieties about the ecological health
and integrity at national *parks*, which generally have had a much
higher profile than historic sites, and which are more likely to be seen
as glimpses of natural ecologies.[21] Still, some site management plans
were quietly revised to foreground (literally) ecological conditions. In
1997, for example, Parks Canada slid the theme of "the relationship
of people and landscape" ahead of the more conventional theme of
"struggle for a continent" at Fort Beauséjour in New Brunswick, explaining that "the surrounding landscape greatly influenced people's
daily lives, and was a major influence in determining where the fort
would be located and the role it played."[22] There was a more nuanced
analysis of the environment's role in explaining the site; it might now
be productive (grassland for cattle at the Bar U, farmed tidal marshes
at Grand Pré), strategic (the view from Fort Edward above the Avon
and St Croix Rivers), or transportational (the shorelines of L'Anse
aux Meadows). As we will see, this proved to be a unique political
moment, when Parks Canada talked in terms of environmental citizenship at historic sites.

But nature has always been an important part of *why* we have considered historic sites important, and especially in North America.
From the nineteenth century, historians and politicians alike equated
natural with national greatness. Nature played a twofold role in New
World history: "her ruggedness was a challenge, and her richness a
manifestation of divine favour," wrote Henry Commager in 1967;
"where in other countries topography is local, in America it is national."[23] Verne Chatelain, the first historian with the National Parks
Service, felt "historic sites would stimulate patriotism in the same way
that scenic ones had for years."[24] Nor has this dissipated in American
mythology; in 2009, Ken Burns deemed national parks "America's
Best Idea" (claiming the entire concept for American ingenuity as well
as the entire continent); three years later, Aaron Sorkin asked in *The
Newsroom*, "So when you ask what makes us the greatest country in
the world, I dunno what the f*ck you're talkin' about. Yosemite?"
Though we deride American jingoism, the sentiment is no less true in

Canada, a country of fewer people and more territory. Histories written through the nineteenth and twentieth centuries emphasized the potential contained in the wilderness and the nation's definition through securing of control over that wilderness. This story of "progress" was the dominant civic narrative, measured by local development and prosperity, community organizations, and improvements in material technology.[25]

Politics in Gingham: Public History and Political Engagement

It is the historian's job to order the past, to make a story out of disparate and sometimes disconnected elements. But public historians – put simply, those that speak to a public audience outside of academia and university classrooms – face particular pressures to craft that story in certain ways. They must negotiate a highly political space as well as a historical one. National parks and national historic sites are designed to promote a feeling of identification with the land, as a source of national identity and a national possession. Consider how historic sites align in a series of "chapters" that together present a story of Canada: eighteenth-century contests to win the continent, nineteenth-century settlement to possess it, twentieth-century development to profit from it. Whether in the cars of the 1967 Confederation Train, the old Canada Hall of the Canadian Museum of Civilization (now the Canadian Museum of History), or what Parks Canada calls the "family" of historic sites, those chapters appear linear and connected. This impulse for linearity affects what we choose to designate and develop, with a teleological emphasis on producing the nation we know today.[26] The concept of family also puts a stamp of belonging on widely dispersed geographies and peoples while implying federal allegiances even at those sites not directly administered by Parks Canada (the extended family).

National history must explain the origins of territorial boundaries, and naturalize those boundaries so that citizens and the rest of the world alike respect those boundaries. It should tout the riches of ecological diversity, but show how such diversity exists harmoniously within a governing political framework. It rationalizes exploitation of natural resources as the basis of our *patrimoine* and collective wealth. And finally, it supplies visual icons by which we know ourselves – think of the view of Lake Louise, or the house at Green Gables. As Benedict Anderson observed, easily reproduced images of landmarks provide the nation-state with "a sort of pictorial census," one of several tools by which to foster an imagined community between people who will never meet face to face, but among whom the state wishes to cultivate a feeling of membership or belonging in a common heritage and territory. In short, the environment is *essential* to this national narrative, as territory, storehouse, and symbol. But there is no political purpose served by drawing attention to the environmental *consequences* of nation building. Instead, the narrative must focus on national growth without cost.[27] While scholars of public history have focused on the civics of commemoration, we need to think about the environmental implications of constructing a national identity and the environmental messages embedded in – or possible for – history education.

Since the 1980s there has been much debate among Canadian historians about how to reconnect with a wider readership, and whether the fragmentation into those limited identities and other subfields effectively "killed" it. Conservative reminiscences have painted a picture of a mid-century golden age when historians (looking exactly as the stereotype suggests) wielded a certain cultural and moral authority, "telling Canadians how the Canadian past ought to have been or in what direction the future ought to go as with dispassionate historical analysis."[28] Ironically, many historians who mourn the loss of such authority also disdain those who actually *do* focus on the public. When I was a graduate student in the public history program at the University of Western Ontario, a professor in the history department tried to get me to rethink my chosen path. "*Why*," he asked, in an accent coloured by a few years at Cambridge, "would you want to spend your career in a *gingham dress* working a *butter churn*?" Clearly, that was not how historians were to tell Canadians anything of substance, let alone how the future ought to look.

Professional bigotries aside, can historic sites challenge the status quo, engage in political debate, or merely reify positive and well-sanctioned stories? Public historians may speak to more people, but they are more constrained than their university counterparts by a wider range of political and material factors, both immediate and ideological. There is the capricious nature of government funding (which has resulted in a problematic emphasis on marketing and sites-for-rent, as we'll see) and the burden of bureaucratic oversight. There is the relative availability of different kinds of artifacts; a disproportionate number of forts and fur trade posts, inherited from earlier generations of designation, compared to stories of the urban poor or immigrant groups or those marginal from central Canada. There is the obligation to cultivate political allegiances and public participation, the result being "heritage": a gentler, more appealing form of history.[29]

And there are the confines of public expectation. The shape of the historic sites system is not simply an imposition of elite conservatism from above, but the product of pervasive conservatism from below. Historical imagery is deeply entrenched in Canadian culture, and with it a deep attachment to certain iconic places and characters. Our expectations of what *looks* historic, or scenic, or Canadian tend to perpetuate a generic quality in historic attractions. As Nicole Neatby and Peter Hodgins observe,

> The use of such formulaic narrative frameworks is often a necessary condition for successful mass communication. In order to make themselves understood, spokespeople for the official culture often have no choice but to draw upon the highly conventionalized words, script, images, narratives, and myths that are culturally available to them and their publics. Furthermore, when faced with a new story, the only means most people have to make sense of it is to draw it back into the circle of their existing cultural expectations, assumptions, and prejudices about how a narrative should be composed, the sorts of characters and plots it should contain, and how it should end. In other words, public communication is conservative in its most profound sense in order to be broadly accessible or legible and believed, the public communicator often has no choice but to work within the restricted vocabulary of the "already-known" and the "commonsensical."[30]

All this to say, I recognize the difficult logistical and ideological position in which public historians at Parks Canada and elsewhere may
find themselves, and that it's far easier for me to propose new stories for
these sites than to overhaul the sites themselves. While museums have
occasionally tested our expectations (often generating firestorms of
protest in the process) and begun to engage in questions of social jus-
tice, historic sites, weighted by the inertia of their physical resources
(much harder to revise than a display case), tend to cater to a safer nostalgia for a preindustrial era.[31] The very boundedness of a site – a parking lot on one side of the visitor centre, cart trails on the other – suggests
a magic door beyond which we may take a brief vacation from our
obligations and concerns in the present, trusting in the known past, the
authenticity of what is in front of us, and the authoritative knowledge
of those who present it to us. "For those willing to experience it," a
former historian from Louisbourg said without a hint of irony, "it is
the true 'magic kingdom,' where time travel is possible."[32]

But site historians know better than most what else can happen
here. Historic sites are places of research as well as teaching, for learning by adults as well as children, and for citizens as well as tourists.
Parks Canada historians have won awards for their work at national
historic sites, showing that first-class research about Canadian history could be done through the prism of a single place. At Louisbourg,
for instance, Anne Marie Jonah linked a colonial outpost to food production and exchange throughout the Atlantic world, while A.J.B.
Johnston begins with the glacial origins of the harbour and ends with
a comment on sea level rise and sea surges attributable to climate
change.[33] Interestingly enough, such engagement has often enough
been rooted in environmental issues. Public history as a recognized
subfield is credited to an American historian serving as a legal wit-
ness for, of all things, water rights in California.[34] Historic sites are
ideally and literally situated to engage in "active history" in the fullest
sense of the phrase, as public education and policy intervention. "The
main thrust of historic preservation today lies not just in educating
those who are already predisposed to appreciating history but in educating those who are not – and who hold the reins of power. Intelli-
gent advocacy groups like historical societies must reach beyond their

own constituency to educate those who are making the decisions which are destroying our natural and man-made environment and causing our aesthetic and cultural deprivation."[35] What one regional planner in California wrote forty years ago amid growing anxiety over suburban build and urban "renewal" projects seems as true today.

Linking Environmental Pasts and Environmental Futures

This book argues that the histories and issues at historic sites can be directly tied to current questions in environmental health and sustainability. In the past, site managers – required to prioritize one period most closely tied to the site's designation – generally opted to artificially maintain a period landscape, which then sat anachronistically in its modern and evolving surroundings. Artificially preventing change has significant ecological consequences; consider the old practice of suppressing forest fires, which we later realized inhibited an essential means of forest renewal. At the same time, aesthetic preferences meant sites were frequently remade with new species that outcompeted native ones. At Grand Pré, for example, an ornamental garden laid out in the 1920s has created an ecological "island" of hundreds of alien plants in the middle of an agricultural landscape.[36] It is a parable about control and the imposition of settler agendas. In other words, these are not islands of history but fragments that tell us about the origins of our current environmental dilemmas – and *why they remain with us*. Making these connections is key. As Robert Melnick, an American landscape architect, puts it, "historic preservation is about the meaning of that past in our contemporary lives; it is about taking explicit human actions to ensure that we do not forget what we did yesterday ... We think so much about history sometimes, about the ways in which landscapes were built and managed and enjoyed and enhanced and even destroyed, that we run the risk of only looking backward ... [We should] make sure that our solutions advance environmental and ecological strength, not merely stabilize it." Historian Rebecca Conard reminds us that environmental problems are fundamentally questions for historians and humanists:

If one believes that, at heart, our environmental problems are really people problems, then it follows that one of the ways to nurture a greater sense of environmental stewardship in the general public is by integrating cultural and natural history. When ordinary people can connect meaning with a place through the observation and contemplation of buildings, structures, and objects – things made by other humans and the history these things embody – there is greater chance they will come to a deeper understanding about the role of human agency in environmental change.[37]

We are still not sure how to do this. Some American scholars credit the National Park Service with having the most experience in, and inclination towards, integrating environmental and cultural resources. But others feel its approach to historic landscapes looks too much like conventional guidelines for built resources: itemizing types of vegetation as "character-defining elements" in the same way architectural features would be; prioritizing historical appearance, design principles, and human use over ecological change, function, or biodiversity.[38] Much the same can be said about Canadian practices. The *Standards and Guidelines for the Conservation of Historic Places*, first issued in 2003, recognized historic sites could range from single buildings to building complexes within land patterns; subsequent revisions (2010) elaborated extensively on cultural landscapes as well as a variety of environmental features, and the concept of change over time at a site. (Both sets of guidelines urge the selection of a singular period of significance for any restoration, for visual cohesion.) Character-defining elements, however, remain key. While the guidelines mandate "understanding the local environmental context" – including climate and winds, geology, topography, vegetation, and ecological processes – there is a notable silence on any issues that may affect the integrity of these elements. For example, the *Standards and Guidelines* warn that the addition of green technologies such as wind turbines or solar panels may affect the visual coherence, and thus the heritage value, of a cultural landscape, but make no mention of why such technologies are now necessary in a world facing the prospect of peak oil, and only refer the reader to British studies about climate change at historic sites. Nor is there any acknowledgment of the number of historic sites – from train stations to industrial plants – that exist because of our historical commitment to nonrenewable fuels.[39]

Despite the enormous growth of environmental history in recent decades, we rarely see references to it in scholarship in public history and almost never in site documents. Management plans, commemorative integrity statements, and system plans treat the environment as the purview of the natural sciences. Even at historic sites, research about past environments is generally done by biologists, bioarchaeologists, or ecologists – and may or may not acknowledge the axis of human/nature interaction that is at the core of environmental history. The *Standards and Guidelines* asks site managers to leave the management of ecological features to "ecologists and other natural resource specialists" and to follow guidelines designed for ecological restoration in natural areas.[40] This is also a reflection of the diminished status of history within the Parks Canada Agency, once (like the National Parks Service in the United States) a leading employer of historians in Canada. But this area of expertise, now recast as "cultural resource management," has shrunk: there are far fewer historians working at Parks Canada, a notable absence of historical memory within the agency, and significant impediments to collaboration between Parks Canada staffers and others. In practice, disciplinary training, the agency's own organizational structure, and the assignment of projects mean that cultural and ecological management remain separate.[41] The material is further divided between sites, since management plans are necessarily directed toward the condition and operation of individual sites, so there has been little in the way of comparative analysis across Canada. All that said, this is *not* a book about policy. The stories here about real places and issues are meant as a deliberate alternative to the jargon of "values management," "stakeholder consultation," and the like – itself a reflection of the growth of managerial, rather than scholarly, capacity in the historic sites system.

I hope to do three things in this book. First, and most directly, I want to script alternative histories for well-known sites, histories that follow people's relationships with nature in the past at these places. Perhaps we can see how our public history might be reimagined as an environmental history, one that acknowledges our enormous debt to the power and fragility of the natural world. Second, I want to use history as a means of integration, spatially and temporally. This book moves across the country, examining regional ecologies in a national framework, relating regional landscapes to national stories. It travels from the medieval Atlantic to the modern prairie, from glacial retreat to global warming. And so lastly, I want us to see these places not as islands of history but as part of a larger habitat of our own making. They should be a part of our public consciousness and conversation about environmental choices, about what kind of relationship we have had and wish to have with our surroundings. Let us ask what we were once, and who we want to be.

CHAPTER 1

Gateway to a New World:
L'Anse aux Meadows

At the northernmost tip of the island of Newfoundland, the shoreline curves into a small cove ringed with rock and a gravel beach. The chill winds of the North Atlantic blow across the beach to a low rise of land, where a cluster of grassy mounds mark the footprint of a group of structures occupied by Norse explorers from Greenland for a few years around 1000 CE. As the only confirmed site of Norse settlement in North America, L'Anse aux Meadows was designated a national historic site in 1968, and the first World Heritage Site in Canada ten years later.[1]

Given its claims of first and only, it is not surprising that L'Anse aux Meadows features so prominently in material promoting provincial tourism as well as the canon of national historic sites. Highly successful advertising campaigns proclaiming Newfoundland "the gateway to the New World" begin here. Canada, too, can use the site to claim a heritage of medieval antiquity far older than any other account of discovery in the settler histories of North America, a story that invokes colourful images of horned-helmeted Vikings and striped sails. More importantly, as a national historic site, it declares a thousand-year presence in, and claim to, the vast waters and near-endless shorelines of the subarctic north.

All this is projected onto what is actually a remote and fairly typical piece of Atlantic shoreline. The site is a more or less empty stage for a millennial mystery. There is just not that much to see at L'Anse aux Meadows, so we must people it either with re-enactors or imaginings. Its environment seems important primarily for ambiance: the chill and fog blowing in from the ocean, the empty beach, the view to the horizon. But charactering the site as a portal to the past through the imagination removes the site, and our experience there, from environmental realities facing Canada's coastal communities that need our attention.

L'Anse aux Meadows speaks directly to – and is literally the product of – three of the most critical environmental issues of our time. For one, it allows us to visualize in concrete terms the effects of long-term climate change. The Norse depended on the warmer temperatures and thinner sea ice of the Medieval Warm Period to sail west from Greenland, and we should recognize that Canada too is profiting from a warming Arctic in expanding its territorial reach. Second, L'Anse marks an early effort to harvest resources from the Atlantic seaboard, a pattern that has persisted for centuries at substantial cost to any number of species, and one that is recorded (but not discussed) in historic sites throughout eastern Canada. And finally, it asks us to consider how and how well small communities that were historically dependent on such industrial harvest are making the transition to a sustainable footing. Despite appearing to be "ripped from the headlines," these all require the long view of environmental history.

"Out to Discover a New World": The Appeal of a Nordic Canada

In 2009, the province of Newfoundland and Labrador aired a series of television advertisements aimed at potential tourists. These were wonderfully done, almost surreal in their vivid colouring and dramatic scenery, whether streetscapes of wooden houses or dizzying views of fjords. One of the most suggestive ads featured a group of fair-haired, cherubic children "discovering" L'Anse aux Meadows. As they prowl wide-eyed around the grassy mounds, spooked yet intrigued by the eeriness of the site, the narrator intones, "And so they came, five centuries before Columbus: fearless warriors, out to discover a new world. The Vikings. While they left behind their mark, they have long since gone ... So far as we can tell." With that, a wooden door swings open, as if by a ghostly hand, and the children flee.

This short clip says several things about what we have come to expect from L'Anse aux Meadows. By describing the visitors as Vikings instead of the more correct but less familiar (and less exciting) term *Norse*, we are to picture helmeted warriors in longships – even if this is not the picture of those who actually lived at L'Anse. (By using the terms interchangeably, Parks Canada is able to cover both bases, speaking with scholarly credibility and popular currency in turn.) The ad, like the site, also tries to make the best of its biggest disadvantage, touristically speaking: the fact that it's a distant, cold, ostensibly empty shoreline. "What they left behind" is actually very little, but the very absence of built artifacts allows for a mysterious, even haunted landscape. And most importantly for Parks Canada and the province alike, the children retracing the footsteps of their millennial forebears invite us, when *we* visit, to reconnect, even reinhabit, an ancient heritage – *our* heritage, and thus *our* claim to the New World. Not bad for thirty-four words in thirty-four seconds.

Of course, Newfoundland and Labrador is not unique in pouring money into heritage-based tourism. The Atlantic provinces have made nostalgia for Maritime heritage and landscapes into a literal art form, as they attempted over the course of the twentieth century to supplement, if not entirely replace, failing resource industries with the more sustainable industry of tourism. But here the Rock has an edge, because it can offer the furthest distance in both space and time. It is nostalgia inflected with the exotic: we can choose between washing snapping on the line by the sea, or a thousand-year Scandinavian outpost. Nor was this new in 2009. Planners for Parks Canada in the mid-1970s envisaged L'Anse aux Meadows invoking same dynamic between the modern tourist and the medieval sailor: "When the decision is made to go to L'Anse aux Meadows, the traveller becomes an explorer on a voyage much like the Norse. The spirit of adventure and discovery, which led to the Norse arrival in North America, are the same elements which must be preserved and presented to the visitor during this travel, arrival and stay at L'Anse aux Meadows National Historic Park."[2] As one scholar has noted, the Vikings of the North Atlantic "epitomize a certain view of people of the North: rugged survivors, predatory, alarmingly charismatic, and shaped by

their challenging climate and surroundings."[3] The key would be pin-
ning such a view and such a people to a particular spot somewhere
in Canada.

Associating the Norse-as-Vikings as part of the "spirit of adventure and discovery" in the story of Canada has been part of Canadian
culture for over a century. In the late nineteenth century, Anglo-
Canadian nationalists liked to position themselves as descendants and,
even more so, inheritors of Norse blood and territory. As Robert Grant
Haliburton famously told a Montreal audience in 1869,

> We are the sons and the heirs of those who have built up a new
> civilization, and though we have emigrated to the Western
> world, we have not left our native land behind, for we are still
> in the North ... and the cold north wind that rocked the cradle
> of our race, still blows through our forests, and breathes the
> spirit of liberty into our hearts, and lends strength and vigor to
> our limbs ... We must be a hardy, a healthy, a virtuous, a daring,
> and if we are worthy of our ancestors, a dominant race. Let us
> then, should we ever become a nation, never forget the land that
> we live in, and the race from which we have sprung ... *We are
> the Northmen of the New World.* Wherever we may go, we shall
> find [that name] "familiar as a household word" and the flag of
> the northmen once more flying upon the ocean, will be a living
> memorial of a glorious past, and the herald of a noble future.[4]

This version of history presented a singular Canadian nation out of a
diverse, divided, and insecure new country, and it gifted that "race"
with the physiological and psychological advantages frequently ascribed to northern peoples in Victorian thought. More importantly
still, it bequeathed to them territorial rights to an uninhabited and
boundless north.

The Norse were a particularly useful point of reference here, both
for their timing ("five centuries before Columbus") and the span of
their empire. Several other places in North America trafficked in Norse
associations in the early part of the twentieth century for the same
reasons. The so-called Yarmouth Runic Stone, for example, prompted
claims that Vinland was actually southern Nova Scotia, and thus

deserving of a Leif Erikson National Park.⁵ But by the 1960s, despite
rival claims and some skepticism, L'Anse aux Meadows had emerged
as the most identifiable site of Norse settlement, thanks in no small
measure to the support of the Canadian government. Nearly a decade
of state-funded archaeology, a willingness to acquire the site as a national historical park, and a formal nomination to UNESCO proposing
L'Anse as the country's first World Heritage Site all signalled Ottawa's
desire to finally run to ground Canada's "Viking" heritage.

Where Was Vinland?

Locating a major historic site on a remote stretch of shoreline on the
northernmost tip of Newfoundland was a project many years in
the making. It was a confluence, as is often the case in history, of
long-term and short-term interests – in this case, of academic curiosity
and political calculation, respectively. By the middle of the nineteenth
century, translations of oral sagas from medieval Iceland reconnected
scholars with the Norse era. Two of the sagas, the *Grænlendinga
Saga* and *Eirik's Saga*, tell how Norse sailors based in southern Greenland discover (largely by accident, blown off-course by storms) a
land further west.

These two sagas describe the features of Vinland and especially the
coastal approach in similar ways, but disagree on who led the expeditions and where the settlement(s) actually was. In the Greenlanders'
saga, Leif Eiríksson sets up camp at a place called Leifsbudir. In the
saga of Erik the Red, Thorfinn Karlsefni and other Greenlanders establish a northerly settlement called Straumfjord and then a southerly
one called Hóp. Both sagas, though, relay the shoreline in three stages:
the Norse encounter first a harsh and barren land they name Hellu-
land (Baffin Island?), then a forested shoreline, Markland (Labrador?),
and finally, "sweet" Vinland. This pattern would have been ideal for
recall in storytelling, but it also provided an oral map of sorts, and
from the nineteenth century, inspired decades of speculation about
where exactly Vinland might be.⁶

In 1960, two Norwegians sailed along the coast of Newfoundland using the sagas as a guide. Helge Ingstad had been an author and

1.1 Parks Canada excavations at L'Anse aux Meadows. Bengt Schonback for
Parks Canada, 1975. (Courtesy of Birgitta Wallace)

administrator in southern Greenland; his wife, Anne Stine Ingstad,
was an archaeologist; both were familiar with the historic Greenlandic
settlements. The Ingstads landed at L'Anse aux Meadows, a tiny outport, where residents pointed out low-lying mounds outside the village.[7] For the next several years, Anne Stine led an archaeological
excavation at the site, which accumulated several Norse-era artifacts
and revitalized the "where is Vinland?" debate in the North American press. After a tepid staff report acknowledged that the site appeared at least to be Norse (or "Viking") in origin, if not definitely
Vinland, the Historic Sites and Monuments Board designated L'Anse
aux Meadows a national historic site in 1968.[8]

Stine Ingstad and Birgitta Wallace resumed fieldwork for Parks
Canada (then the Canadian Park Service) between 1973 and 1976,
during which time Prime Minister Pierre Trudeau formally opened
L'Anse aux Meadows National Park. Shortly thereafter, the International Council on Monuments and Sites (ICOMOS) pronounced

L'Anse aux Meadows "an outstanding property of Man's heritage" and the first World Heritage Site. "It is," said the World Heritage Committee, "a precious and, up until now, unique milestone in the history of human migration and the discovery of the universe."[9] The next year, Parks Canada began reconstructing the three halls that would become the signature part of the site. Living history programs and costumed interpreters followed in the early 1990s, and annual visitation rose from a few thousand to over twenty-four thousand in 2012–13.[10]

The halls and the larger anthropocentric emphasis on "Man's heritage" are understandable for a cultural or historic site. But they are also misleading. Archaeological reports and interpretative plans from the 1960s and 1970s constantly referenced the environmental context, millennial and contemporary. Researchers from the Ingstads on recognized the natural features that would have enticed the Norse to land here, including a beach, a brook for fresh water, and a peat bog with raw iron ore. In 1972, an international advisory committee was tasked specifically with researching the physical environment of AD 1000, and ICOMOS would reference paleoecological as well as archaeological evidence as justification for World Heritage status. At the same time, many of these same features posed any number of difficulties for the archaeologists, who had to deal with snow melt and spring flooding, the instability and acidity of the sphagnum and peat bogs, and a very short field season.[11]

"The Very Atmosphere of the Place": Ambiance as Interpretation

Even with the reconstruction and visitor centre completed in 1985, L'Anse aux Meadows was not a typical historic site. There was (and is), quite simply, very little *historic* for visitors to see. The visitor centre presents a handful of artifacts and a scale model of the Norse site, a film, and interpretive panels that treat Norse culture and its archaeological recovery outside. Visitors then exit to gaze out over what looks to be a fairly ordinary, fairly bleak stretch of north Atlantic shoreline – what ICOMOS described bluntly as the "unastonishing appearance of its archaeological vestiges" – before descending to the boardwalk and crossing the bog to the reconstructed sod halls. The public experience of L'Anse aux Meadows relies heavily on ambiance instead of artifact, on the natural setting instead of material remains, on the sensuality of a landscape with the suggestion of history rather than a historical structure set in a particular place. The Ingstads described being captivated by "the very atmosphere of the place, its situation and landscape";[12] the historic site needs visitors to be similarly enthralled so they can accept the lack of relics and be willing to *imagine* what happened here, to experience history in their mind's eye.

This message has pervaded writing about the site for forty years. "You come to L'Anse aux Meadows not so much to see, for the site is not spectacular – but to think," said the *Montreal Gazette*. The *Canadian Geographical Journal* was typically moody: "Looking across Épaves Bay from L'Anse aux Meadows it is easy to imagine the arrival of Norse ships 1000 years ago. With the existing shelters removed, the Norse building outlines turfed over, and all traces of the 20th Century cleared away, the visitor will again be able to appreciate the stark, quiet loneliness in which these few traces of the final westward expansion of Norse culture have lain for almost 1000 years." The *Globe and Mail* went further: "As you approach the bleak seacoast, you shiver in anticipation. Chill, barren isolation is the perfect setting for what you are about to experience – a journey back through time … You stand there wrapped in thought and fog on the edge of the Atlantic, imagining the longboats slicing through the waves, square wool sails bellying with the wind as they sail away into the unknown."[13] And this was the exact premise of "And so they came" in 2009. The absence of artifacts becomes a virtue; the empty and apparently unaltered shoreline an evocative backdrop for projecting oneself back in time. The stark and undeveloped shoreline also ensured a kind of heritage purity; the site would never become "Disneyland Norse," promised the *Globe and Mail*, an unsubtle swipe at a more commercial and less natural kind of destination south of the border.[14]

The empty shoreline also framed the central and perennial question of L'Anse aux Meadows: *was* this Vinland? This allowed Parks Canada to stickhandle around the scholarly debate (by posing it as a question while simultaneously claiming to be the only viable candidate), and it gave the site a quality of mystery almost unique in the historic sites system. The National Film Board titled its 1984 account of the Ingstads' discovery *The Vinland Mystery*; twenty years later, the award-winning Canadian Mysteries website asked "whether this tiny isolated hamlet, clinging for survival between a hostile, often stormy, sea and a barren swampy hinterland, was the fabled wine-producing Vinland of the sagas."[15] David Blackwood featured a giant Viking ghost floating above a sod hall in his 1985 print *L'Anse aux Meadows* for the Historic Sites Association of Newfoundland and Labrador;[16] in 1992, one of the wildly popular Heritage Minutes called "Vikings" featured the Norse as ghosts, transparent against the landscape, vanishing into the distance.[17] The shiver that went down your back ... was that merely the chill of the wind off the ocean?

From the outset, interpretation at L'Anse aux Meadows was designed around its capacity for ambiance. A visit would be a dramatic and sensory experience instead of an intellectual exercise or lesson to be learned; interpretation would take "an evocative rather than a purely didactic approach," using "intangible elements of feeling and association."[18] But these intangible elements were grounded, quite literally, in tangible ones: the physical features of the site. In drawing the site's boundaries, the view out to sea was deemed essential for understanding what the coast might have looked like for those arriving from Greenland a thousand years ago. As a result, 60 percent of the site (about fifty of its eighty square kilometres) is actually marine, encompassing thirty-four kilometres of coastline plus inshore waters and islands. Although most visitors troop east from the interpretative centre directly to the archaeological terrace, a hiking trail loops through bogs and along the shoreline, offering a "landscape little changed since Viking time."[19] Just like the children, exploring as "fearless warriors," visitors are to empathize *with* the Norse more than learn *about* them.

In its emphasis on the imaginative and the intangible, L'Anse reflected the new thinking about historic sites in the mid-1970s. The excavation and reconstruction of the Fortress of Louisbourg on Cape Breton continued apace, the largest historic reconstruction in Canadian history. But Louisbourg was now an anomaly, and not just because of its size. Parks Canada was increasingly unwilling and unable to tackle fixed reconstructions with living history programs, preferring instead to cultivate "cultural landscapes" that offered a "sense of place" and to discuss successive waves of occupation over time. This evolution is particularly noticeable in Atlantic Canada, with its long history of historic sites and designations. The eighteenth-century Fort Anne at Annapolis Royal, one of the first national historic sites, had been partly reconstructed in the 1930s, while across the river Samuel de Champlain's *habitation* at Port Royal was *entirely* (and not entirely accurately) rebuilt. By the 1970s, in contrast, sites like L'Anse aux Meadows and Port au Choix in Newfoundland and Melanson Settlement in Nova Scotia consisted of quiet, grassed-over archaeological sites. Visitor centres were designed to blend into the background, with site-sympathetic architecture corresponding to local topography, just as the grey concrete at L'Anse aux Meadows bulges like granite outcropping over the bog.[20]

There is something to be said for a more passive approach to historic interpretation. It frees the visitor (and the researcher) to think creatively, rather than follow a single storyline, and it allows the setting – the context for human action – to make its presence felt. That said, it is debatable how effective ambiance alone actually is in educating the public. The decision to add costumed interpreters at L'Anse by 2000 suggests it is not.[21] Just as the image of Viking marauders in longships has overshadowed the utilitarian and prosaic reality of the Norse settlers, a handful of grassed-over mounds cannot compete for attention with the reconstructed dwellings and the costumed storytellers, and don't work nearly as well in advertising. Conveying information is simply easier with material evidence and human actors, especially if we are trying to understand a distant past and a way of life further removed from our own. This is particularly true for national sites, whose higher profile will draw visitors from further afield, from a greater variety of backgrounds, and with a wider range of knowledge (or lack thereof) about the site. In short, ambiance can be powerful, but as a teaching tool, it has its limits.

"The Situation of the Land Was Beautiful": Finding Vinland

Thanks to these two major trends of the 1970s – the archaeological turn, and thinking about cultural landscape – L'Anse aux Meadows was suffused with environmental content from the outset. Several exhibits in the visitor centre allude to the natural or geographical context of the Norse world. There is a discussion of Norse seafaring and navigation. There is some reference to the natural resources that would have attracted the Norse and subsequently ensured their survival, and the theory – advanced by Birgitta Wallace – that Vinland actually refers to the entire mouth of the Gulf of the St Lawrence, with L'Anse aux Meadows serving as a way station or base camp for expeditions further south. And finally, there is the archaeological evidence that the Norse smelted natural iron ore from Black Duck Brook for making nails to repair their ships.

All that said, the environmental context is a supporting player in the national narrative here. The reconstruction suggests a residential settlement rather than a way station: a permanence that supports a Canadian claim to the land more than the very brief occupation of a few years by the Norse. And nature is presented an instrument for human ingenuity. In a typical example, the text panel on smelting states, "[Here the Norse] had the natural resources, the ability, and the requirements of survival." This is very much in keeping with Parks Canada's larger story about building a nation through human endeavor and innovation.[22]

But L'Anse aux Meadows is somewhat unusual among Canada's historic sites because much of its original research and planning consisted of paleoecological, rather than historical, studies. This was partly due to the fact the site had so few manmade features contained in a significant amount of territory, but it may also reflect new anxieties emerging in the 1960s about the wear and tear on national parks and their ecological integrity. As one report began, "having responsibility for the preservation and protection of parks and historic sites for future generations in an unimpaired condition, requires a management plan based on comprehensive knowledge of all biotic and physical components of the park's environment."[23] Between 1974 and

1976 Parks Canada, together with the Geological Survey of Canada and the Canadian Forestry Service, assessed numerous aspects of the L'Anse aux Meadows site: land (bedrock geology, topography, soils and drainage systems), water (coastline, drainage), and wildlife populations and vegetation (forest zones and plant communities). Some of this involved slotting the area into categories established by the new national land inventory, but much of the research consisted of studies of wood, peat, bog iron, and pollen samples dating from the Norse era.[24] This was more historical ecology than environmental history in that it was primarily concerned with measuring the non-human by species, separate from human history, and in the distant past, but it did help construct the ecosystem and climate of a thousand years before.

But historians were thinking about the environmental context of Norse exploration. The central preoccupation with Vinland has always been the question of its location, so scholars measured the sagas' descriptions against the Canadian coastline, trying to retrace Norse sailing routes based on place names, landmarks, currents and winds, and what landscape characteristics would most appeal to settlers from Greenland. That last is particularly important to the current management plan, which emphasizes "a relatively undisturbed natural landscape whose features and resources led the Vikings to establish their base camp here."[25] This refers to both the resources on site and within sailing distance, because of the theory – most consistently argued by Parks Canada archaeologist Birgitta Wallace – that "Vinland" refers to the Gulf of the St Lawrence, as an entire region from which the Norse harvested such things as wood and grapes. (Grapes, obviously, don't grow in northern Newfoundland, neither in medieval or modern warming periods.)

For decades, then, scholarship about L'Anse aux Meadows has intersected with questions about the environment and the environmental past. But it has not engaged with environmental history as such – that is, the relationships between human ideas and actions and the natural environment. More crucially, the site can use what we learn about these relationships to shed light on some of the most pressing environmental issues facing us in the twenty-first century. How might it do that?

A New Narrative for L'Anse aux Meadows

A Country of No Winter: A Longer View of Climate Change

It is difficult to visit L'Anse aux Meadows and *not* think about climate. Standing on the ancient beach facing out to sea you might picture a Norse sail in the distance. But you will likely see chunks of ice floating down "iceberg alley," as the province (which promotes iceberg-watching as a tourist activity) calls the current carrying ice floes south. Now you might picture an Arctic ice shelf somewhere to the north calving into open water, and satellite photographs showing the Northwest Passage open to a degree unprecedented in recorded history. This is one of the few places that we can actually see global climate change – arguably *the* environmental issue of the twenty-first century, and the one with potentially the greatest impact on Canada's place in the world – at work.

Parks Canada has acknowledged the implications of climate change at a few historic sites, though only briefly. It identifies two particular concerns: the melting of permafrost under northern sites such as the stone Prince of Wales Fort at Churchill, and the increase in storm surges and shoreline erosion at coastal sites like the Fortress of Louisbourg. Text panels at L'Anse aux Meadows do mention periods of warming and cooling like the Little Ice Age. More revealing, though, is the fact that the only reference to climate in the current management plan is to the "climate of limited financial resources."[26] Given the number of historic sites in coastal locations or located on major rivers already prone to flooding (I think particularly of the Forks at Winnipeg), and the acceleration of designations in the North, it is time to make climate a central part of understanding our past.

Confronting the causes and effects of changing climatic conditions at historic sites has several advantages, beyond the inescapable and pragmatic managerial concerns about site integrity. It rounds out our sense of "what life was like," and provides an empathetic connection to our predecessors who had their own anxieties and concerns about global weathers. It questions the conventional practice of recreating a steady historic state, when in fact climate changed during the past that we attempt to revisit.[27] Perhaps most importantly, it presents an issue usually couched almost exclusively in scientific language in human terms. It acknowledges the inescapable relationship between climate and human action, whether as part of a context shaping human opportunity or part of the global ecosystem affected by human actions.

In this sense, L'Anse aux Meadows *is* well suited for a "journey back through time." The beach terrace is tangible evidence that coastlines are not fixed or permanent geographies. As the last glaciers retreated about twelve thousand years ago, their meltwaters flooded islands in the North Atlantic into underwater hummocks, drawing centuries of fishing fleets to what would become known as the Grand Banks. Meanwhile, the Great Northern Peninsula, released from the weight of the ice, began to rise out of the ocean in a phenomenon known as isotatic rebound. It is still rising. The beach where the Norse camped a thousand years ago is now about four metres above sea level and one hundred metres inland, and one of three stepped terraces that mark the slow lifting of land from sea.[28]

We know that the Norse would have encountered warmer temperatures – indeed, this was how they were able to reach the New World at all. Archaeologists at nearby Port au Choix, on the Strait of Belle Isle, have found that fluctuating temperatures in seawater affected the movements of marine mammals and the Indigenous peoples (Groswater and Dorset Paleoeskimo) who pursued them.[29] In the same way, Norse voyages from Iceland and Greenland were made possible by a period of warming in the North Atlantic with a decline in sea ice. We know there have been a series of long-term climactic oscillations stretching back thousands of years, alternately warming and cooling in periods of five hundred to eight hundred years. The best-known example is the Little Ice Age (approximately AD 1300–1850), when cooler weather beset the agricultural economies of early modern Europe, but there is also a substantial body of work on the phase that preceded it, the Medieval Warm Period. This warming enabled the expansion of the Norse sailing and farming in northern altitudes from the tenth through the fourteenth century (when the cooler temperatures of the Little Ice Age again constricted growing seasons and contributed to the collapse of the Newfoundland and Greenlandic settlements).[30] The sagas tell us that a large part of Vinland's allure was its mild temperatures, evidenced by the fabled grapes and, critically for the Greenlanders dependent on cattle, winter foraging, "The country seemed to them so kind that no winter fodder would be needed for livestock: there was never any frost all winter and the grass hardly withered at all."[31]

We need to recognize ourselves in this longer history, because Canada, too, has moved to claim new territories when the climate allowed. The country elbowed its way onto the continental plains in the mid-nineteenth century, using a framework of agricultural settlement to claim vast amounts of the interior. Such political ambitions needed to cast the desired territories as desirable environments, or ones that would support farming. It helped, then, that Canada was sending scientific expeditions and political feelers out onto the prairies in the 1850s, or the end of the Little Ice Age, when the prairies were cooler and wetter, and thus greener. This meant surveyors (and boosters) encountered – using the same language as a sailor approaching the shores of a new land of riches – an "ocean of grass" and "seas of verdure."[32] (It would take a half-century of unsustainable expansion of intensive farming and a catastrophic drought to erode such meteorologically misplaced overconfidence.)

Parks Canada acquired L'Anse aux Meadows in the late 1960s, when Ottawa was increasingly intent on claiming the Arctic archipelago. In fact, Canada had been using environmentally based rationales for territorial claims in the North for over a century: through a racialist rhetoric of "northerness," and scientific expeditions and North West Mounted Police patrols designed to both fly the flag and identify any available resources.[33] With its transnational (or more precisely, prenational) history, Vinland touched a competitive nerve among northern states eyeing Arctic territory at the height of the Cold War. The *Ottawa Citizen* urged the federal government to take "more than an academic interest" in the site and fund the excavations in order to beat out "the modern Norse" (referring to the Norwegian Ingstad and Danish researchers). After all, the newspaper argued, the sagas mention a child born in Vinland, which made him "in a manner of speaking the first native-born Canadian." Americans were loath to give up the possibility that Vinland was located in New England, a claim sustained by Congress' decision in 1964 to declare 9 October Leif Erikson Day. After decades of American traffic on purported

Canadian territory, from pipelines to air defence, Ottawa was keen to publically identify the North as Canadian.[34]

Equating ecological and territorial integrity helped this campaign significantly. When the ss *Manhattan*, an American oil tanker, became the first commercial vessel to cross the Northwest Passage in 1969, Parliament responded with a law (the Arctic Waters Pollution Prevention Act) claiming jurisdiction over all waters within a hundred miles from Canada's coastline and allowing Canada to regulate any shipping in these waters. Prime Minister Pierre Trudeau announced that "Canada regards herself as responsible to all mankind for the peculiar ecological balance that now exists so precariously in the water, ice and land areas of the Arctic Archipelago."[35] As a legislative mandate, it was admirable, if ambitious; as a political strategy, it was shrewd, asserting national authority in the name of global ecosystems, invoking a selfless ecological rationale to defend jurisdictional self-interest.

Meanwhile, after four decades of no new national parks, the federal government began creating new parks in spades and with particular speed in the North. There was also a concerted interest in northern history. Parks Canada developed Dawson City (of the Klondike gold rush) in the Yukon, and referred to L'Anse aux Meadows as having an "arctic feel" and "near-Arctic character."[36] It also won World Heritage designation for Nahanni National Park in 1978 – making Canada's first two World Heritage Sites definitively northern sites. "World heritage sites" are something of a misnomer; they must be nominated, and managed, by national agencies, and so reflect much more directly the values of the nominating country than the international community. The period of Norse occupation is so distant from the history of Canada as a nation-state as to make the site essentially anachronistic to the national narrative in *actual* terms, but not in symbolic ones.

It is doubly ironic that a nation of southerners invokes the Norse to claim its Arctic affinity, when both their era and ours – the ends of each millennium – are periods of warming. Then as now, we move into the North when we can, when it seems *less* like the North. Canada's scientific, military, and legal claims to the Arctic seabed and northern waters have escalated to an unprecedented degree in the past decade, just as the Northwest Passage appears to have opened permanently. In 2007, the Conservative government established a perennial multi-pronged presence (military, scientific, industrial) in the Far North. The centerpiece was the highly publicized (if questionably named) Operation NANOOK, a military exercise designed to troop the flag in an Arctic that is ever more marine and more trafficked. Tellingly, NANOOK is also for practicing responses to environmental disasters such as oil spills.[37] In 2013, Canada submitted to the United Nations Law of the Sea a massive new claim to northern territory: another 1.75 million square kilometres in the form of continental seabed. This would redraw its borders and extend its territorial reach for the first time since 1949, and with public support; polls show that a majority of Canadians see Arctic sovereignty as the country's top foreign-policy priority, more so than peacekeeping. All of this, of course, is because of the new opportunities in a melting Arctic for accessing undersea oil and gas deposits: the finite and thus infinitely valuable riches of the industrial age.[38] And most recently, the search for – and discovery of – John Franklin's ships in 2014 and 2016 by Parks Canada was appropriated so quickly and thoroughly by the federal government that it was difficult to see even the underwater archaeological excavation of mid-nineteenth-century ships as anything but a statement of contemporary territorial claims to a northern sea-passage.[39]

So L'Anse aux Meadows represents a gateway to a new world for the medieval Norse and modern Canadians alike. The historical parallel is just too tempting. Finding ancestors in another sea-going and shipbuilding power that expanded its commercial empire in search of natural resources in a time of Atlantic warming is a subtle way of naturalizing our current national ambitions. (L'Anse aux Meadows served as a repair station for oceangoing vessels; one of the major components of Canada's Arctic strategy is building a new refueling station in Nunavut to service the new icebreakers and northern patrol ships included in a massive shipbuilding project.)[40] But such a heroic narrative does not acknowledge the climactic context that makes such ambition possible, the warmer temperatures that have made the Arctic accessible. More importantly, it does not address our culpability in helping to create that context. The warmer temperatures during the so-called Anthropocene are fundamentally unlike those of the Medieval Warm Period, in that the current climate is the result of unsustainable human action: industrialization and the large-scale production of greenhouse gases. Where we *can* learn from the Norse is to take caution. They expanded into new places (Greenland, Vinland) during warmer conditions, but their reach contracted when temperatures cooled. They were, in effect, trying to adapt to two new environments – one geographical, one climactic – and their existing environmental knowledge did not help them adapt to climactic instability.[41] We too are expanding into an unfamiliar environment (the Arctic) at a time of climactic instability (the Anthropocene), neither of which have much reference point in our normal vocabulary. Pride goeth.

"Much Valuable Produce": The Exploitation of Marine Resources

Just as we pursue oil and minerals off our Atlantic coasts, the Norse were drawn westward in search of desired commodities. There was an aesthetic value to the new land; the sagas describe the new country as beautiful, and, as we've seen, gentle and mild, the dew "the sweetest thing they had ever tasted." But the greatest impression of Vinland is the accounting of unprecedented riches. Some of these would sustain the settlers themselves: "They made use of all the natural resources of the country that were available, grapes and game of all kinds and other produce." Most, though, were to be harvested and transported back to Greenland: "They made ready for the voyage [back to Greenland] and took with them much valuable produce, vines and grapes and pelts." Not surprisingly, the sagas tell us, "these expeditions were considered a good source of fame and fortune."[42] As one scholar put it rather dramatically, "paradise was glimpsed, at least, if not actually found, somewhere in the golden west. And with it came the promise of untold wealth by exploiting the abundant natural resources of the land they had found."[43]

L'Anse represents the promise of that wealth, the gateway to paradise. The site itself was valuable for its landing beach and bog iron, but it also gave access to resources further afield. Birgitta Wallace has argued that L'Anse aux Meadows served as a base camp for explorations south into the Gulf of the St Lawrence. Here a different ecological zone, the Acadian forest, would have offered warmer conditions for the mysterious wheat and grapes mentioned in the sagas, and pieces of butternut found at L'Anse aux Meadows.[44] There is also evidence that the Greenlanders traded with the Dorset and Thule Inuit peoples of the Eastern Arctic, especially in pursuit of walrus ivory. This would have given the Norse access to extensive networks of exchange reaching from Labrador to the High Arctic at Baffin Bay and the Davis Strait.[45]

L'Anse aux Meadows has been presented as the opening chapter in the story of Canada, the first of several moments of discovery, exploration, and contact between Europeans and First Nations. And it does represent an early example of a quintessential national story – but *not* that of settlement. L'Anse did not mark the permanent arrival of Europeans; it shows their *im*permanence, and the reason why. As Wallace explains, "the purpose of the expeditions was clearly stated to be a search for resources such as lumber and other things that could give riches and fame, with no attempt at permanent settlement. In all instances, the return to Greenland was taken as a given."[46] As a short-lived outpost in a resource hinterland, L'Anse aux Meadows models one of the most fundamental of Canadian patterns, what historian William Morton characterized as the "alternate penetration of the wilderness and return to civilization [as] the basic rhythm of Canadian life." (Morton cited Vinland as an early example of staple harvesting.)[47] In exporting assorted renewable resources from the western Atlantic – carting biomass from the New World to the Old – the Norse foreshadowed the European and then the Canadian agenda in the New World, and its waters, for a thousand years to come.

Perhaps even more importantly, the story of L'Anse aux Meadows reminds us that the Norse were not alone in the "new" world. Scholars have focused on a cooling climate as the reason for the Norse retreating back across the Atlantic; the onset of the Little Ice Age by the mid-fifteenth century made it too difficult to maintain outposts on exposed shorelines thousands of miles from Iceland, let alone Norway. But in history there are usually forces pushing *and* pulling at our human actors. A narrative of heroic arrival and mysterious departure by the Norse obscures the influence of the indigenous people who greeted them. If colder weather – building sea ice, shrinking the seasonal windows for sailing and pasture – pulled the Norse back over several decades, the sagas suggest that they were also pushed off the shores of Newfoundland by resident peoples (who may have been ancestors to the Inuit, Beothuk, or Mi'kmaq, depending on when and where in the sagas we are). In these stories the "Skraelings" are alternately useful trading partners, especially for furs – "In exchange for the cloth they traded grey pelts" – and armed groups that overpower the Norse and send them back to Greenland. In *Eirik's Saga*, "Karlsefni and his men had realized by now that although the land was excellent they could never live there in safety or freedom from fear, because of the native inhabitants." Even the Norse-authored sagas betray two fundamental flaws in their attempt to colonize this new frontier: the strength of Indigenous occupation, and the fact that the Norse attempted, unsuccessfully, to manipulate the exchange. At different points, they cheat their new trading partners or attempt to extort goods through violence. Thus, if their impermanence was partly intentional – a resource outpost – it was also partly unintentional, frozen out by climactic and social factors.[48]

Yet this pattern of incursion, extraction, conflict, and retreat is not one discussed at Canada's historic sites. The public face of the national sites system in Atlantic Canada is the "battle for a continent" at eighteenth-century forts such as Halifax and Louisburg. But there is a remarkable cluster of sites spanning several centuries and coastlines that remind us that rival empires actually had very similar interests. Consider L'Anse (tenth-century Norse); Red Bay (sixteenth-century Basque whaling); Canso, Ryan Premises, Lunenburg, and Battle Harbour (French, English, and Canadian fisheries, from the fifteenth to the twentieth century). It is telling that three of Canada's seventeen World Heritage Sites are drawn from this group – the greatest concentration on a single theme. Historians and historical ecologists are now sketching the ecological character of the early modern North Atlantic, finding evidence of overharvesting and habitat degradation on European inshore waters. The loss of these fisheries coincided with, and indeed, propelled exploration westward, as fishing fleets chased species across the Atlantic. Suddenly the medieval Atlantic is not that far away from us:

While the Atlantic islands' environmental legacy of the Viking Age introduction of farming includes severe erosion and widespread degradation of soils and vegetation, it appears that these chiefly agricultural societies were often effective long-term managers of marine resources. However, in the broader historical view it has recently become clear that the Norse in the North Atlantic bear the mixed responsibility for the origins of western European commercial fisheries that have so drastically altered marine ecosystems worldwide over the past two centuries.[49]

Of course, the Norse never dreamed of harvests on the scale carried out by the fleets that converged on the Grand Banks or Georges Banks centuries later. But the Norse did establish a pattern of colonization and extraction that would be repeated, again and again, by Europeans and then Canadians – and which would be normalized and heroized in Canadian history just as it had been in the sagas.

To its credit, Parks Canada nominated Lunenburg and Red Bay for World Heritage designation partly through Criterion V, which recognizes "an outstanding example of a traditional human settlement, land-use, or sea-use which is representative of a culture (or cultures), or *human interaction with the environment especially when it has become vulnerable under the impact of irreversible change.*" But we have shied away from the real implications of this history, and its lessons for us today. Quite understandably, we prefer stories of triumph to tragedy: the start of an era rather than its end, the discovery of bounty rather than its exhaustion, examples of human innovation and ingenuity rather than of profligacy and profit. For example, Lunenburg focuses on the romance of the "saga of the sea" and the age of sail, not the unsustainable nature of the industrial fishery. In a profound irony (or a most cynical political maneuver), the town was nominated for historic site status at precisely the same time as a federal moratorium closed the Atlantic groundfishery: making it both a living, and silent, testimonial to the collapse of the fish stocks in the early 1990s.[50]

ICOMOS *declined* to recognize Red Bay under Criterion V, despite acknowledging that a relatively short Basque occupation had indelible effects on whaling populations. Parks Canada's choice of emphasis

may have influenced this, but it certainly accords with it. While Red
Bay acknowledges the ecological cost, the site emphasizes "the adaptations of the 16th-Century Basques to the harsh marine and terrestrial environment of Labrador as they became world leaders in the
hunting of whales and the processing of whale oil." In other words,
crediting physical resilience, economically driven determination, and
technological ingenuity as helping the Basque profit from a new environment – a variation on the Norse, having "the natural resources,
the ability, and the requirements of survival." Parks Canada also lists
the successive fleets (Basque, French, and English) that harvested numerous marine species (fish, whales, seals) over centuries, implying
perennial wealth in the waters that would become Canadian.[51]

But this allows us to continue to think of these waters as bountiful,
not finite. We need to see the historic sites in Atlantic Canada as part
of another millennial pattern, a pattern of seasonally and systemati-
cally harvesting a series of marine species to exhaustion. Site interpretation disconnects the cumulative effect of centuries of industrial
harvest from current practices – a quintessentially *ahistorical* view of
our place in nature. A more engaged and applicable reading of Red
Bay, for instance, would be as oil to oil: showing the parallels between
the hunt for whale oil and the search for fossil fuels. Whale oil and
whale products fuelled, literally, certain items of modernity that we
take absolutely for granted: light, heat, lubricants, plastics. In the mid-
nineteenth century, the discovery of petroleum allowed us to substi-
tute one type of oil for another without disrupting the supply of
consumer goods to which we had become accustomed.

Red Bay is an opportunity to talk about how much we rely on oil-
based products, how these are supplied, and their sustainability – elements missing from the public discussion about Canada's oil industry.
Canadian energy producers depict a heroic story of innovation and
"made in Canada" ingenuity while quietly reinforcing the myth of in-
exhaustibility and new frontiers in a language of exploration. Oil roy-
alties remain the key to empire building among Canadian provinces,
including Newfoundland and Nova Scotia. But just as Basque sailors
endured a "harsh marine and terrestrial environment" in pursuit of
whales, the hunt for offshore oil and gas hinges on a careful balance
between the potential for profit and that for disaster.[52] As historian

Paul Sabin has argued, a historical perspective reminds us that we have substituted older energy systems for new energy sources before. While our reliance on fossil fuels is enormous and unprecedented, a transition to new (i.e., renewable) sources is still within the realm of possibility.[53]

Along the Viking Trail: The Past and Future of Coastal Communities

People living along the Great Northern Peninsula do not need to be reminded of the difficulty of adjusting to a new way of life when confronted with the end of an older one. After two wrenching interventions in the past sixty years – resettling outport communities and closing the groundfishery – they now find themselves living along "the Viking Trail," nearly five hundred kilometres of highway that brings tourists from around the world to a rebranded peninsula promising dramatic coastal scenery and medieval heritage. The trail ultimately deposits visitors at L'Anse aux Meadows National Historic Site, within sight of the village of the same name. While the two appear to be (deliberately) isolated from one another, their fates are linked. The historic site that commemorates a Norse community is also meant to sustain a Newfoundland one, "maintaining the prospect for a long-term tourist industry in the area."[54]

In a country still dominated by extractive industry, Atlantic Canada offers the most advanced example of what happens when this kind of economy has run its course. The region has already experienced the end of the industrial lifespan, in overharvest and collapse. For most of the last century, it has used heritage tourism in an effort to graft a different and relatively sustainable means of revenue to counter or at least slow its declining status in the Canadian federation and the global economy. This was particularly true in the postwar decades of big government when the Atlantic provinces and the federal government actively promoted historic reconstructions in remote and economically struggling communities. Such was the thinking behind the largest reconstruction in Canada's history, located in a place that illustrated most poignantly the limits of the resource economy: the Fortress of Louisbourg, amid the shuttered coal mines of Cape Breton. The same applied to fishing communities.[55]

1.2 View of L'Anse aux Meadows, national historic site and village, 2009. (C. Campbell)

It is a good strategy in many ways, and it appears to be successful. By 1980, the highway freshly paved, L'Anse aux Meadows drew ten thousand visitors to "this particular piece of hell-and-gone Newfoundland," as the *Montreal Gazette* rather uncharitably put it.[56] Recent exit surveys show the Great Northern Peninsula is the most popular region to visit after the capital of St John's, and visitors here are nearly twice as likely to visit a national park and national historic site. In the past few years, L'Anse aux Meadows has attracted upwards of twenty-five thousand visitors – far outpacing any other national historic site in the province. This is particularly important in a region that has experienced dramatically higher rates of population decline and unemployment compared to the rest of the province.[57]

Heritage tourism does make history (in some form) more prominent in public culture, and it also serves as a tacit acknowledgment that we can, in fact, exhaust our natural resources. But there have been unintended consequences. The historic site is often physically and socially detached from the surrounding communities, as at Louisbourg and L'Anse aux Meadows. This may be the result of genuinely different geographies, past and present (in both these cases, the nineteenth-century settlements were located near to but not on top of the older ones), as well as the need to protect the fiction of living history sites of travelling back in time. But it can also establish an awkward if not outright hostile dynamic between come-from-away historic staff and local residents, instead of the *noblesse oblige* symbiosis imagined by bureaucrats.[58]

If, on the other hand, the local population does embrace the period identity chosen for it, we can see too much enthusiasm, specifically for its marketing potential. The historic event grounded at L'Anse aux Meadows has been diffused over for five hundred kilometres, and over a range of commercial ventures all tagged with the Viking theme (including a Viking dinner theatre in what promises to be the only sod-covered restaurant in North America).[59] Across the road from the historic site sits Norstead, a fictional Viking port built in 2000 by the Viking Trail Tourism Association as a living history project and funded by both the federal and provincial governments. Its activities and costumed interpreters are far more extensive than at L'Anse aux Meadows, but the most authentic thing about Norstead is the replica *knarr* that sailed from Greenland to L'Anse aux Meadows in 1998. While this regional marketing may show local buy-in, or at the very least a pragmatic business sense, one scholar has argued that it is more insidious: a neo-conservative effort to devolve responsibility for economic development to the community level.[60]

My concern is that thinking of heritage tourism as a post-industrial strategy allows us to believe we are already, safely, post-industrial when our economy and ethos remains strongly rooted in the resource industries. The ad campaigns celebrate the traditional in some cultural practices (fiddling, washing on the line, children playing), but not all: they excise any reference to extractive industry, even though this is the strongest and most vibrant tradition in Atlantic Canada. That the historic site of L'Anse aux Meadows is within sight of the fishing port of the same name suggests a passing of the torch, a gentle evolution from declining industrial harvest to sustainable service

economy. But it can be read in exactly the opposite way: the presence
of the modern village demonstrates the resilience of the industrial era.
Beak Point, which extends out from the archaeological site, was to
have been acquired by Parks Canada some time after 1975, but remains in provincial and private hands. A small commercial fishery
still operates here. The forest line has been cut back over the past two
centuries and is now fifteen kilometres away.[61] Newfoundland and
Labrador, like all the provinces, is still tumbling after revenues from
the energy sector. Gros Morne National Park has had its status as a
World Heritage Site repeatedly threatened by proposals for hydro
transmission lines running through the park and for fracking nearby.[62]
A replica furnace hut opened at L'Anse aux Meadows in 2000 was
co-sponsored by Petro-Canada and Norsk Hydro Canada Oil & Gas,
both significant leaseholders in offshore exploration and partners in
projects like Hibernia. Portraying the site as a "proto-industrial"–
that is, site of the first iron smelted in North America – is plausible
enough. But in light of the province's current agenda, it seems frankly
opportunistic, the very definition of a usable past: looking for historic
precedent to justify present actions. The hut telegraphs corporate efforts to buy goodwill, not a reasonable precedent. Offshore oil plat-
forms are a different – it must be said – kettle of fish than medieval
boat repair.[63]

Designation as a historic site should not excise a place from contemporary uses, politics, or values. The language of a "family" of
national historic sites suggests a peaceful and unifying presence
of federal administration in a provincial landscape, but land is and
has always been the greatest flashpoint of conflict between the two
levels of government. Historic sites are still property, and extractive
industry continues to define eastern Canada. Parks Canada has little
or no leverage in the tug-of-war between Ottawa and St John's, and
the agency doesn't license fishing boats or oil wells, but it does profit
from place. Since the early part of the twentieth century, it has justified its existence by pointing to the tourism revenue it generates at national parks and sites, as well as its role in providing for Canadians'
"benefit, education and enjoyment" by administering these places.
In recent years, significant budget cuts have inspired a consistently if

somewhat cheesy entrepreneurial turn, whether sanctioning corporate sponsorship (see above) or inventing new activities (with additional fees) for visitors.[64]

The Viking Trail, and L'Anse aux Meadows, demonstrates how essential land and landscape is to profitable heritage tourism.[65] The trail
succeeds not in spite of its remoteness but because of it: a rugged
"wilderness" geography and a linear route to the end of the world (at
least Canada's), just as the Norse sailed to the furthest reaches of their
maritime world. The scenery is both antimodern and ahistorical, with
the "raw glory of the Newfoundland wilderness, the Viking scale of
its fjords" reminiscent of that earlier age yet perennially available for
our visual appetites.[66] Little wonder the views from the beach at L'Anse
aux Meadows out to sea and across adjacent-but-not-yet-designated
lands are so important. "Vinland the Good" is the ultimate in desirable real estate: connotations of respite and rural bounty for medieval
sailors, and as a preserve of wildness for us.

L'Anse as Historical Park: A New/Old Kind of Site?

Landscape has always been key to the meaning and value of L'Anse
aux Meadows as a historic site. As it has been to the historic sites system as a whole from its very inception, beginning with the cultivation
of the Plains of Abraham as a historic landmark in 1908. J.B. Harkin,
the first director of the Dominion Parks Branch, envisaged a category
of historic parks as a means of shoehorning recreational space (and
the Branch's authority) into the more populated eastern provinces, by
demarcating green space around historic ruins. In the mid-twentieth
century, the Branch targeted historic sites with large and complex properties as both symbols of region within the national narrative and
engines of economic development. These included the Fortress of Louisbourg, Rocky Mountain House, Batoche, and L'Anse aux Meadows.[67]
At such historic parks, situation and setting clearly added to the story,
but also supplied space for outdoor activities. Early plans for L'Anse
envisioned "opportunities for extensive recreation experience, *par-
tially or wholly independent of the Norse theme*, in order to provide the

visitor with a more complete day of activities." Initially the discussion *so* emphasized outdoor recreation that some felt obliged to caution that it "should not be allowed to overshadow the Norse interpretation."[68]

Ultimately, this would not be a concern; as we've seen, the costumed interpreters at the reconstructed hall are a much bigger draw than the hiking trail. What is interesting, though, is how the setting inflected "the Norse interpretation" right from the start. Planners and researchers alike understood Norse history to be entwined with natural history: "the two should, to some extent, complement each other"; "both land and marine components [should be seen] as the greater natural backdrop, support area and buffer of the Norse site"; "the theme ... will be 'people and nature' rather than nature alone."[69] Many of the ideas for exhibits took their cue from the Ingstads' on-the-ground (or -water) perspective: the attractions of Atlantic Canada for Norse explorers, their navigation technology, gauging the location of Vinland by sailing times and shorelines. It sounds very much like what we think of as environmental history – that is, the interrelationship between human action and the natural world.

A second theme in the original plans seems equally prescient. There was a particular interest in succession, connecting the various groups that have occupied the site, "how it has changed since 1000 CE – if at all – and how the various settlers have used its resources." There is a remarkably pervasive argument for continuity, or more precisely commonality, between settler cultures who have shared "a basic dependence for survival on the same set of natural resources." The original development concept stated quite forcefully, "The Park as a whole will be used to illustrate ... the relationship and dependence of the Norse settlement and more contemporary cultures on the natural resources of the Park."[70]

Drawing this parallel now would be politically daring, because it implicates our own economic agenda, but it would also be ethically responsible. In a rather disappointing contrast, the current management plan (from 2003) locates, and contains, resource use and dependency in the distant past: "Visitors will understand the Viking's [*sic*] focus on exploration and exploitation of resources, and the challenges they faced as strangers in an already inhabited land." Exploitation is presented as a historic phenomenon, not a current one. The "traditional harvesting activities" that are acknowledged and accommodated on-site, such as hay gathering and berry picking, are limited, harmless, almost bucolic.[71] This, to be blunt, lets the rest of us off the hook.

Could the concept of historic park be revived, and stretched, to engage us beyond scenery? Could we recast the beach at L'Anse aux Meadows to discuss our historic dependence and impact on the North Atlantic? Or could it be the first of a new kind of marine conservation area, one that focuses on human history on and around the ocean? Marine conservation areas are the next frontier for Parks Canada and UNESCO, but to date they have been defined largely if not exclusively by ecological criteria alone.[72] Would this allow Parks Canada to discuss ecological and commemorative integrity in a holistic way, and more importantly, to show us how current environmental issues are rooted in past actions? I know this is ambitious; we are not accustomed to seeing our historic sites as places to question ourselves and our ideas about Canada. But there is something about the gateway to the New World that entices us to think beyond the horizon. History is not meant to be merely an affirmation of who we are or have become. Environmental history, especially, asks us to acknowledge that our choices have consequences and costs.

CHAPTER 2

Idyll and Industry: Grand Pré

In June 2012, UNESCO named the landscape of Grand Pré a World Heritage Site as "exceptional testimony to a traditional farming settlement created in the seventeenth century by the Acadians in a coastal zone with tides that are among the highest in the world." Using dykes and *aboiteaux*, the Acadians reclaimed salt marshes along the Minas Basin in the Bay of Fundy to produce nutrient-rich farmland (the *pré*, or meadow) that is still drained and farmed today. To UNESCO, then, Grand Pré's universal heritage is primarily as an artifact of early European coastal land management and a living tradition of agricultural practice, as well as a memorial to the Acadian deportation (*le grand dérangement*) by the British before the Seven Years' War. The same is true of its designation as a national historic site and Canada's first rural national historic district.[1]

By drawing a direct lineage between the pré's creation and current practice, between Acadian settlers and "their modern successors," the UNESCO designation permits and even encourages us to vault over the intervening era of industrial agriculture. There are certainly interesting and valuable continuities, such as the complementary use of uplands and marsh/dykelands by farming families, to the collective management of the pré by a community-run marsh body. But this stewardship narrative linking Acadian and modern farming also perpetuates a rural idyll, an image of the place that has migrated easily from an older romantic nostalgia to the newer language of environmental and community sustainability. While this suits the interests of both Acadian cultural nationalism and provincial tourism, it excises a kind of land use that bears far more directly on land use today, and that formed the basis of Nova Scotia's economy and identity for much of the nineteenth and twentieth centuries. Grand Pré demonstrates how a place can be made iconic as non-industrial, even as it participated in a global industrial economy.

Grand Pré sits at the northern entrance of the Annapolis Valley, a stretch of flat land about 130 kilometres long, between the North and South Mountains – a relative use of the term, since the "mountains" are well under three hundred metres high, but still high enough to be significant in the low-lying Maritime provinces. In a province with only 8 percent land in agriculture,[2] the valley's sheltered microclimate and the silted shores of the Minas Basin have been especially valuable, creating an area with the most abundant and diverse agricultural production in Nova Scotia. But if the pré proper is the result of seventeenth-century dyking, its agricultural geography in the Annapolis Valley is at least as much or much more directly the result of *nineteenth*-century ideas about agriculture, the environment, and the state. This chapter, then, proposes a second narrative for Grand Pré, one that treats the site as part of an Anglo-Canadian Annapolis Valley as well as *l'Acadie*, and that recognizes the continuity of the industrial past as well as the continuity of dykeland farming. The challenge is to recognize the importance and the rarity (if not the uniqueness) of the pré as a seventeenth-century artifact operating in twenty-first-century time, while showing how it has been affected by subsequent decisions and historical patterns. To integrate these two strands is to present to the public a more complete history, a more expansive geography, and a more useful discussion of agricultural sustainability.

Grand Pré figures prominently in the history of colonial power and commemorative practice in Canada. Historians have concentrated on its place in eighteenth-century Acadie, and its construction as a site of romantic pastoralism for the tourist gaze in the twentieth.[3] But its environmental history can draw from other disciplines, notably a long tradition of historical geography in the Maritimes, and recent archaeology focused on the Acadians' salt-marsh farming. Historians of the early modern Atlantic world have found similar marshland practices in New England and northwest France.[4] Research on the history of Nova Scotia agriculture (including apples, the province's most iconic cultivar) acknowledges the effect of climate and soil, but has concentrated on economic output rather than environmental circumstance or impact. I suspect this is because of a preoccupation with the political economy of regionalism, and the longstanding – if not central – debate as to whether Nova Scotia was made poorer or richer by Confederation with Canada.[5] So nature has always been present here, but in the background or in neighbouring disciplines. We have yet to script a history of environmental change at Grand Pré, and consider what it means for the shape of agriculture in Canada today.

"The Fruitful Valley": Grand Pré in the Annapolis Valley

In the Acadian land, on the shores of the Basin of Minas,
Distant, secluded, still, the little village of Grand-Pré
Lay in the fruitful valley. Vast meadows stretched to the eastward,
Giving the village its name, and pasture to flocks without number.
Dikes, that the hands of the farmers had raised with labour incessant,
Shut out the turbulent tides; but at stated seasons the flood-gates
Opened, and welcomed the sea to wander at will o'er the meadows.
West and south there were fields of flax, and orchards and cornfields
Spreading afar and unfenced o'er the plain.
– Henry Wadsworth Longfellow, *Evangeline, A Tale of Acadie*

Henry Wadsworth Longfellow was not a saltmarsh biologist, nor did he ever actually visit "the shores of the Basin of Minas" – but his famous 1847 poem *Evangeline* got the basics of Acadian coastal agriculture more or less correct.[6] The pré was created when, beginning about 1680, French Acadian settlers drained the salt marshes between the mainland, Long Island, and Boot Island to create a roughly circular area of farmland now measuring about 1,300 hectares. Anchored by marsh hay and cordgrass, fed by the silt of the Fundy tides and the wide, slow-moving Cornwallis and Gaspereau Rivers, and sheltered from Atlantic weather by the uplands to the southeast, the area boasted all the advantages that much of the rest of Nova Scotia, with its oceanic exposure and slate bedrock, lacks.[7] Indeed, it has been suggested that the Edenic description of the New World and the references to fruit in the tenth-century Icelandic sagas in fact point to a location in southern Nova Scotia for the fabled Norse settlement of Vinland.[8]

The largest Acadian settlements, including Grand Pré, Beaubassin, and Port Royal, lined the eastern Fundy shore with farmland and pasture. *Aboiteaux* drained seawater out of the marshland into solid pastureland, while dykes – which did require "labour incessant" in construction and maintenance – kept the tidal flow at bay. The British eyed these rich farmlands even before the French surrendered mainland Nova Scotia in 1713, but especially after the British founded a new capital at Halifax in 1749. Situated on a world-class harbour but a decidedly unarable peninsula of slate and granite, the planted town and naval base required settlements of supply. The British thus looked to the Fundy shore and particularly Grand Pré in the Minas Basin. "This place might be made the Granary not only of this Province but also of the neighbouring Governments," wrote Paul Mascarene in 1720.[9] Anxious about their hold on the former-and-still francophone Acadie, and envious of the natural wealth of the Fundy area, the British spent several decades trying to secure the potentially hostile yet desirable Acadian lands. After brutally deporting thousands of Acadians beginning in 1755, they encouraged waves of Anglo-American immigrants to settle in the Annapolis Valley. Planters from New England in the 1760s, and then Loyalists after the Revolutionary War, arrived to find rich farmland literally ready-made. In this, the *dérangement* applied with terrible and focused efficiency the same imperial strategy of displacement and resettlement – of plants and peoples – that reshaped the entire hemisphere.

The newcomers used these lands in much the same way as the Acadians had: relying on the complementarity of uplands and marsh, locating woodlots and residences on the uplands and agriculture and pasture on the pré, mixed farming with a heavy emphasis on livestock. Unlike the Acadians, though, they enjoyed a certain amount of state support, established a system of land division and town planning that persists to this day, and could count on emerging colonial markets for their produce.[10] Such was the rationale for building the Shubenacadie Canal, which would cut through the colony to link the Bay of Fundy to the Atlantic shore. The canal was proposed as early as the 1790s and begun in the 1820s, to supply Halifax with the produce of "our finest, best cultivated, and wealthiest agricultural districts."[11] It would become the longest canal in Maritime Canada, a permanent reminder of the wealth and value of Fundy farmland to the rest of Nova Scotia.

That value would only increase over the next century. If geography asks us to think in terms of scale, and history asks us to think in terms of time, then the story of Grand Pré needs to expand outward to include its regional context, and successive chapters in its history. Scholars have recognized it as part of a much larger imperial geography, but it is also part of an industrial one. If the physical pré worked by farmers in the twenty-first century is the product of Acadian expertise and labour in the eighteenth, then the cultural, political, and technological landscapes it occupies are generally the result of nineteenth-century Anglo-American ideas and practices that drew the pré into much larger networks of production. We can draw a direct line from today's site to the seventeenth century, but that line runs through the ideologies and practices of modern agriculture taking shape in the intervening years.

Consider the moral standing that family agriculture held in the Anglo-Atlantic world through the eighteenth and nineteenth centuries. Accepting his nomination to represent Hants County, just east of Grand Pré, in the Nova Scotia colonial assembly in 1863, William D. Lawrence proclaimed, "I am more in favour of agriculture than any other [industry], the first great employment of man – the noblest employment of man – agriculture, which takes one from his fireside, into the fields where with the plough he turns the soil to the face of heaven, and casts the seed on with his hands, and waits with patience to a kind providence for the reward of his labour."[12] That Lawrence was a shipowner and shipbuilder, in a colony at the peak of its merchant trade and shipbuilding industries, speaks volumes about how much weight voters placed on the social and symbolic value of farming. Little wonder that colonial and subsequently Canadian authorities would present it as the preferred form of land use, as the key to economic self-sufficiency and social stability. Historians writing through the twentieth century ascribed similar qualities to the Planters, as a kind of genteel pioneers, with a "love of the soil, sobriety, industry, and thrift" in "planting well."[13]

The story of rural progress, prosperity, and civility was institu-
tionalized in the provincial museum system in 1971, when the Nova
Scotia government acquired Acacia Grove at Starr's Point across the
Cornwallis River from Grand Pré. This was the stately Georgian home
(built 1814–16) of Charles Prescott, a shipping merchant from Halifax who moved to the Annapolis Valley to take up a new career as a
scientifically minded gentleman farmer. While he experimented with
hundreds of fruit species, Prescott is known primarily for introducing
some key commercial apple varieties, notably the Gravenstein, which
became the province's signature product. Prescott House linked an
architecture of wealth and individual success (a common feature of
local museums) to the new landscape of orchards and "the development of the apple industry and its role in the local and provincial
economy from the 19th century into the modern period."[14]

Today Acacia Grove attracts a tiny fraction of the attention given
to Grand Pré, but its commemoration was certainly deserved in the
larger history of Nova Scotia. By the middle part of the nineteenth
century, the orchards at Starr's Point had become the international
image of the colony. When Nova Scotia apples won medals at the
Royal Horticultural Society's International Fruit and Vegetable Show
in 1862, it elevated the colony's profile within the Empire and
affirmed the prestige of the agricultural sector. As R.G. Haliburton,
serving as secretary for the colony's commission for international
exhibitions, proclaimed to a loyal audience,

> The agricultural capacities of the province are, I believe, un-
> surpassed. The alluvial lands of the Bay of Fundy are without a
> parallel in the history of agriculture ... Nowhere can a farmer,
> with so small an amount of skill and industry, make so com-
> fortable a living as in Nova Scotia ... And Nova Scotia, hitherto
> supposed to be only callable of rearing fir trees, has sent some of
> the best oats in the Exhibition; and it has been actually proposed
> that that land of perpetual fogs should send home a cargo of
> oats, to be used as seed by the British farmers. Then, our apples
> and potatoes sent there, [sic] are almost unrivalled. What could
> we not do if we could only import a few Mechis and model
> farms to the shores of Minas Basin, and give our province the

same advantages which those have enjoyed that have competed with us at the World's Fair?[15]

Nova Scotia would dine out on its standing as the "Orchard of the Empire" for the next eighty years, enjoying the symbolic, political, and economic currency of its special relationship with, especially, British markets. Botanical display and exchange was fundamental to the imperial project, of course, but also to Nova Scotia's view of its place within that Empire, given the global traffic of its fleet. (This also fuelled much of the colony's opposition to Confederation with the Canadas in the 1860s: with the ties to empire visible daily in her ports, it seemed unnecessary if not counterintuitive to "turn their backs upon England and fix their thoughts upon Ottawa.")[16] While her sea captains brought home plants as biological trophies, symbols of their connections in distant ports, the Annapolis Valley was reshaped for commercial export, to compete with other agricultural producers around the world. The Shubenacadie Canal was operating by 1856, but much more significant, especially for transporting perishables, was the network of railways remaking the political and industrial land-scape of British North America. Numerous lines – lined with new warehouses for produce storage – connected Valley farms to inland hubs such as Truro and ocean ports such as Halifax.[17]

After Confederation in 1867, Nova Scotia found itself drawn into a second new network: a national program of agricultural science. In 1886, the new federal Department of Agriculture established half a dozen Dominion Experimental Farms across the country, including one at Nappan, at the head of the Bay of Fundy. The Experimental Farms were designed to test and distribute crops that would fare best in different regional ecosystems; they thus encouraged specialization and concentration, whether Marquis wheat in Saskatchewan or par-ticular apple varieties in the Annapolis Valley. (The long-term effect is brought home in the popular memoir/mandate *The 100 Mile Diet: A Year of Local Eating*, when J.B. MacKinnon realizes that the handful of apple varieties in a grocery store represent a mere fraction of the dozens that once were grown in British Columbia: a diversity now ut-terly forgotten.)[18]

The province was equally invested in cultivating expertise. A School of Agriculture was established at the Provincial Normal School in Truro in 1885, and a School of Horticulture at Wolfville in 1905. Significantly, there was also a vibrant culture of what Parks Canada now calls "citizen science" at the local level. In January 1910, for example, the *Berwick Register* reported that the Farmers' Meeting ("well attended, and proved very interesting") heard a series of reports ("at length and most instructively") on methods of packing and grading, new forms of disease, fertilization, and transport and reception in London.[19] That year, at the urging of the Nova Scotia Fruit Growers Association, Ottawa agreed to establish an Experimental Fruit Station at Kentville. The Kentville station remains a dominant presence in Valley agriculture. It was responsible for introducing new varieties that quickly became provincial signatures, like the Honeycrisp apple and *l'Acadie blanc* grape. And it gamely hosts the annual Kentville Pumpkin Festival to lure tourists in the fall.[20]

Indeed, tourism has been intertwined with farming in the Valley since the late nineteenth century. The same railway networks that helped industrialize agricultural production promised tourists from New England a romantic, Arcadian landscape of prosperous farms wreathed in blossom. The Dominion Atlantic Railway, which purchased land at Grand Pré in 1917, landscaped the site (complete with a statue of the fictional Evangeline) as an ornamental flower garden visibly distinct from the working fields around.[21] But selling the Land of Evangeline meant selling the *land* as much as the Evangeline. Ambling by automobile in 1923, Charles Towne grew impatient with "a certain lithograph which had greeted us from almost every inn at which we stopped, showing Longfellow's heroine standing in the midst of a radiant orchard, got a bit on our nerves. I was weary of the vicarious glimpses of apple-trees, and I knew that if we didn't find them soon in their bright abundance we should feel cheated." Travellers at the Grand Pré station paused for a moment of hushed regret for the poor, departed Acadians, but then resumed their rhapsodic descriptions of "seas of bloom," and their praise for the rich soils and evident prosperity of the Valley. "Of Grand Pré it has been said it boasts a three-fold attraction: beauty, fertility, and sentiment," explained the Canadian

2.1 Farm and view across valley, Richard McCully Aerial Photograph
Collection, 1931. (Courtesy of the Public Archives of Nova Scotia)

Pacific Railway. "No wonder the Acadians were blythe," mused one
1894 tourbook, "this must have been a veritable land of plenty." And
clearly, it still was: "Acadie, home of the happy," then and now.[22]

Such a garden view offered a perfect complement to the other historic landscape in the region. The eighteenth-century star-shaped forts
scattered around the Fundy shores (Fort Anne, Fort Edward, Fort
Beauséjour, Fort Gaspareaux) represented masculine ambition and
imperial conflict, in fixed and geometric ways. Grand Pré, despite its
tragic history, now suggested peace, not war. Its references were feminine, emotional, and organic, a picture of annual renewal rather than
historic ruin. At the same time, it was a reassuringly *productive* space,

where grazing cattle and apple blossoms meant profit in the present as well as evocations of the past. This was a profitable pastoral, a new symbolic landscape seemingly immune to the fluctuations and conflict in the industrial economies of fish or coal.[23]

The promotion of agriculture as "beauty, fertility, and sentiment" persisted through the twentieth century. Valley farmers launched the Apple Blossom Festival in 1933 to promote their exports and import the tourists. (Recognizing, as others had done with Evangeline, the allure of the apple blossom as a picture of feminine innocence, the festival annually crowns a Queen Annapolisa from among Apple Princesses representing the Valley communities; the first Queen was crowned at the Kentville research station, thereby uniting photogenic ceremony with practical interest.) Soon after the province themed "scenic travelways" to promote different microregions, with the blossom-ridden Evangeline Route – perhaps not coincidentally, provincial Highway 1 – a calming counterpart to the "coastal sublime" of the Cabot Trail on Cape Breton; the regions, and their matching highways, remain the backbone of provincial tourism today. The Valley as living pastoral blended into an appealing kind of folk culture. Stan Rogers sang the lament of the expatriate, remembering the view of the Minas Basin centred on Grand Pré:

I wish I grew Annapolis apples up above Fundy Bay
Oh, it seems so far away
On the ridge above Acadia's town to the valley down below ...
Down on the farm, back among the family, away from Ontario
Hear the ladies singing to the men, dancing it heel and toe
And watching the apples grow ...
I've climbed the ridge of Gaspereaux [sic] Mountain, looking
   to the valley below
And watching the apples grow.[24]

Now the agriculture-themed celebrations aimed at city-dwellers from Halifax (or Ontario, or Boston) fill half a calendar year: the Apple Blossom Festival in early June; Tastes of the Valley and Meet Your Farmer Days in harvest season along with county and provincial exhibitions; a month-long wine festival in September, and the Pumpkin

Festival in October.[25] Farmers from the Valley also maintain a year-round presence in Halifax, through community-supported agriculture subscriptions, farmers' markets, and a new group of locally supplied grocers. Even Parks Canada is now cultivating an apple orchard on the historic site next to the church at Grand Pré.

Commemoration, Sustainability, and *Terroir*

Given the agricultural history written onto this landscape, and its long-standing public appeal, it is not surprising that Grand Pré has been presented, physically and metaphorically, as a garden. But the growing popularity of organic and locavore food movements calls us to look more closely at their evolution, and the industrial practices to which they are such a deliberate response. The current marketing of Grand Pré as a model of sustainable agriculture is both new and familiar. The image is one of aesthetically pleasing, small-scale farming, still pinned to the dykeland, but adopting the newer language of sustainability and *terroir*. Wineries, in particular, have proliferated in the Gaspereau area since the 1990s, touting distinct grape varieties and vineyard practices geared to the coastal soils and climate, and attempting to gain an international reputation in language that recalls that of apple producers a century earlier. "Tidal Bay brilliantly reflects its birthplace: the terroir, coastal breezes, cooler climate and our winemakers' world-class craftsmanship," proclaims Domaine de Grand Pré, housed in an 1826 farmhouse.[26] Even non-agricultural products made elsewhere in Nova Scotia trade on the Valley imprint of "lush fields and orchards ... and the harmony man shares with nature."[27] These messages of local adaptation, regional character, food security, and environmental empathy are then wrapped in the final sanction of Acadian precedent as "people who nearly perfected the art of land cultivation and sustainable agriculture."[28]

The commemorative language used to designate and promote Grand Pré as a historic site encourages these associations. The site first was revised to present Acadian occupation, past and present. In the early 1980s, the older scripts from the Historic Sites and Monuments Board and Parks Canada, emphasizing the Deportation and the

2.2 View of Grand Pré to Cape Blomidon, 2012. (C. Campbell)

Seven Years' War, gave way to language that celebrates Acadian settlement before 1755 *and* the emotional "attachment [that] remains
to this day among Acadians throughout the world to this area, the
heart of their ancestral homeland and symbol of the ties which unite
them."[29] With the move away from a two-nations narrative in Canadian historiography, Parks Canada made recognizing other "cultural
communities" as a strategic priority in commemoration. At the same
time, the Acadian community became more and more invested in the

management of the site; Grand Pré is now jointly managed by Parks Canada and the Société Promotion Grand-Pré. So the dykeland, for example, is framed as a substantial work of coastal engineering and a national achievement – that is, of the Acadian nation.

The new emphases on the Acadian community and on landscape in site commemoration reinforce a direct relationship between the Acadian past and current practice. Grand Pré's designation as a national rural heritage district cited "the blending of natural and built features, *in the retention and development of land use patterns originating with the Acadians*, particularly in the spatial distribution of arable land, orchards, dykelands, and residential hamlets."[30] The designations also draw a direct line between the dykeland's management by preindustrial and contemporary community groups, between the collective effort of the Acadian settlers and the marsh bodies of today. In fact, retaining traditional lands, uses, and management are all key to sustaining the place itself. The dykes must be maintained to physically protect the pré from tidal flows; the pré must be farmed to maintain a continuity of function and meaning; and it must be locally managed to preserve a sense of community autonomy and local knowledge.

This is where the elasticity of the term *sustainability* comes in handy, as it weaves between historic and ecological meaning. Certain farming practices have *sustained* this historic pré over time; the result may be a kind of agriculture that is more *sustainable* in terms of natural resources. Other agricultural districts in northwestern Europe and the Northeastern United States have, like Grand Pré, "cultivated" an image of a pastoral to diverge intentionally from large-scale, scientifically managed production.[31] Certainly, UNESCO makes this connection at numerous agricultural landscapes around the world:

> Cultural landscapes often reflect specific techniques of sustainable land-use, considering the characteristics and limits of the natural environment they are established in, and a specific spiritual relation to nature. Protection of cultural landscapes can contribute to modern techniques of sustainable land-use and can maintain or enhance natural values in the landscape. The continued existence of traditional forms of land-use supports biological diversity in many regions of the world. The protection

of traditional cultural landscapes is therefore helpful in maintaining biological diversity.[32]

Presenting Acadian land use patterns as effective in their permanence and sustainable in their footprint benefits Parks Canada (in presenting itself as an agent of ecological integrity and environmental education), the Acadian community (supporting their historic claim to the area with evidence of good environmental stewardship), local farmers (as the caretakers of this tradition and an alternative to harmful industrial practices), and Canadians (who can feel good about a fragment of community agriculture surviving into the present day).

However, it perpetuates a Golden-Age characterization of Acadie, which severely limits our understanding of the site in both space and time. As Graeme Wynn has argued, we can credit the Acadians with knowledgeable affection and respect for the Fundy marshlands. But European colonization was fuelled by a confidence in human mastery of nature, and the Acadians, like most settlers, had a pragmatic, functional view of the land and its potential yield:

> Much as we may wish to see them as such, neither the early indigenous peoples of the region nor pre-expulsion Acadians can properly be counted as proto-environmentalists. We can allow them concern about the lives and livelihoods, the well-being, of their children, and even their children's children, but there is no evidence that they appreciated the biophysical limits of their settings or what we would now call ecological linkages in anything other than purely local and practical ways ... We should not mistake these images as evidence of deliberately forward-looking, ecologically aware, sustainable practice.[33]

And yet, such is our investment in the Grand Pré pastoral that it persists. Paintings and murals commissioned for the site show *l'ancienne Acadie* as a "quasi-paradise" in perennial harvest, with "wonderful weather, abundant agricultural productivity, and a happy and carefree existence."[34] When I taught in the College of Sustainability at Dalhousie University, even my Nova Scotian students, who have been schooled in Acadian history since elementary school, wanted to see it as evidence of a once and future moment of sustainable agriculture, a wooden-shoed carbon footprint for the modern age.

Meanwhile, the alluring image of farms in harmony with nature only serves to increase external pressures on the area: much like it did in the railway age, but now encouraged further by a feel-good stamp of sustainability. The Annapolis Valley is the most important liaison between the urban and rural in Nova Scotia, and Grand Pré is both the gateway to and face of that connection. The mutual dependency of the Atlantic and Fundy shores has existed since the eighteenth century, but the twinning of Highway 1 now underway will make "our finest, best cultivated, and wealthiest agricultural districts" even more accessible. Indeed, one of the main reasons for rebuilding and maintaining the dykes in the mid-twentieth century was to support the highways now so essential to both agriculture and tourism.

In this relationship, the pré's collectivist, nonindustrial face is its fortune. Accordingly, there is no mention of the active, if not more problematic, history that was required to create this landscape, whether biological (new species introduction and specialization), chemical (sprays and fertilizers), or material (the physical infrastructure of production and transport). The Annapolis Valley was not exempted from the arrival of industrial agriculture; indeed, as the province's key agricultural region, such practices would have been concentrated here. Nor is there public discussion of the economic health of the agricultural sector over the course of the twentieth century: the postwar loss of international markets, the growth of multinational agribusiness, farm consolidation, and rural outmigration.[35] In this chapter of its agricultural history, Grand Pré once again represents the larger Maritime region. Parks Canada's concern for ecological integrity and the campaign for UNESCO designation have revealed external ecological pressures on the pré, including a preponderance of non-native species, rising sea levels and tidal pressure on the dykelands, and other land-use development proposals in the area. The 2011 landscape management plan acknowledged that conservation of the property requires a working agricultural economy, and that the problems facing the farmers on the dykelands "are not unique to Grand Pré."[36] This is perhaps the frankest acknowledgment that the pré cannot be seen as separate from the larger economic and ecological life of the Valley.

Nor is the pré separate from transatlantic patterns of exchange and displacement over centuries. Grand Pré is the product of a migration of peoples and ideas about coastal settlement, a migration that defined the early modern Atlantic world. Whether in imperial conflicts of the eighteenth century or the imperial commerce of the nineteenth, land was an invaluable commodity for enhancing state power, promoting agricultural settlement as a mark of social stability and prosperity, creating peaceful images of prosperous occupation, or satisfying new consumer preferences. Grand Pré demonstrates how a place's cultural identity can be scripted externally yet prove indelible from and unexpendable to the place in the long term. From its relationship with a famous poem to its designation as a World Heritage Site, the place is granted meaning from away. A working agricultural landscape cannot be isolated from its surrounding environment – indeed, historically, it would not have been.

Of course, the idealized, singular periodization we see at Grand Pré appears at historic sites across in Atlantic Canada. The public image of Prince Edward Island is also of *its* rural golden age, in the latter nineteenth century, when the Island was one of the most agriculturally productive districts in Atlantic Canada. The Island idyll, in television, tourism, and Cavendish National Historic Site, presents an era when thousands of small farms embodied the social coherence and stewarding sensibility of the family farm and the economic sustainability of mixed farming.[37] Back in Nova Scotia, Lunenburg does the same with the Grand Banks fishery of the same period, as a kind of proto-industrial, community-knit enterprise of low-impact technologies, and sustainable abundance reflected in the prosperity of the high Victorian streetscape. The heroics of daring sea captains and the romance of the "saga of the sea" are untroubled with any reference to the groundfish moratorium, let alone the larger history of resource harvesting in the North Atlantic or the socio-political debates surrounding the fishery today.[38] While public history generally requires both a positive story and a clear-cut message, environmental history tends to muddy those waters.

This comes back to the value of Grand Pré's history – if taken altogether. As a model for small-scale, local, and low-impact farming, it has the very real effect of affirming the art of the possible, of humanizing

and cultivating support for sustainable agriculture. The site is a wonderful starting point for thinking about succession and alternative types
of land use. But it is not a straight line between the seventeenth century
and today. It needs to be understood in *relational* terms: the Acadian
past in relation to what came after, Grand Pré in relation to the rest of
the Annapolis Valley and the province, and today's dykelands in relation to industry norms. The idyll is most useful, in other words, when
contrasted with what followed – when it is shown as an alternative to
mainstream agricultural practices. But that means we must discuss
those mainstream practices, a reality in which Grand Pré also exists.
Even with the concrete artifact of the pré, a strong Acadian voice and
sense of history in the community, and a growing interest in sustainable practices among Valley farmers, most of the frameworks from the
nineteenth and twentieth centuries for industrial agriculture remain in
place, frameworks of science, technique, infrastructure, and identity.
Until we incorporate this second story, we will continue to see Grand
Pré simply as "Acadie, home of the happy."

CHAPTER 3

Wilderness, Lost and Found: Fort William

Located on the northwest coast of Lake Superior, at the mouth of the Kaministiquia River, Fort William was one of the most important fur trade posts in North American history. From 1802 to 1821, it was the central depot for the North West Company (NWC), the dynamic rival of the Hudson's Bay Company (HBC). Each summer it hosted a huge *rendez-vous* when voyageurs from Montreal met their counterparts from the interior districts, each travelling as far as the open water would let them, to exchange trade goods for furs. As the fur trade faded from the northwest, the post was surrounded by the rail yards and grain elevators of the Canadian Pacific Railway (CPR). From a storytelling perspective, Fort William had it all: nation-making exploration, colourful multicultural trade activity, and a political contest of wills. The problem was that as a historic site, Fort William had nothing *left*.

In 1971, the Ontario government announced the largest historical reconstruction in its history: rebuilding Fort William upriver on the Kaministiquia. Today, Fort William Historical Park appears to be an unqualified success. It is the largest reconstructed fur trade post and one of the most complete representations of fur trade life in the world. It attracts up to one hundred thousand visitors a year, making it one of the most prominent tourist destinations in the vast expanse of Northern Ontario. But there are instructive stories from the early years of the reconstruction project that speak to significant dilemmas in public and environmental history. The province contracted the reconstruction to a private company, National Heritage Limited, whose focus on a quick reconstruction and a profitable operation sparked heated exchanges with archaeologists and historians, who complained that, "profit will carry the field over authenticity."[1] While other major state-funded reconstructions in Canada – including the Fortress of Louisbourg, Fort Edmonton, and Sainte-Marie among the Hurons – wrestled with similar questions, Fort William represents an important experiment in privatizing public history, and the compromises and constraints facing public historians.

The design of the new Fort William also tells us a great deal about expectations of the environment in Canadian history. While archaeologists continued excavating the site of the original fort in the CPR rail yards, the reconstruction took place some fifteen kilometres up the Kaministiquia River at Pointe de Meuron. The new site was different from the original in important ways, but it held one trump card. Upriver the reconstruction would be surrounded by woods; at the lakeshore, it would be surrounded by a rusting city core. No convincing fur trade post could rise again in such a setting; popular ideas of the fur trade in Canadian history – spanning and securing the future territory of Canada – demanded at least the suggestion of a vast boreal forest. Even today, the fort promotes the "wilderness" setting of its campgrounds and conference facilities. And the long-standing demand for an unproblematic wilderness setting cloaks several pressing environmental questions about our use of that same boreal landscape.

Fort William, 1802–1923: The Northwest Entrepôt

While the Hudson's Bay Company accessed the interior of the continent from that "frozen sea" from 1670, its rivals travelled west through the Great Lakes and inland rivers – what Washington Irving lyrically described as a "singular and beautiful system of internal seas."[2] French traders began using the Kaministiquia as early as the 1670s as a means to Rainy Lake, Lake of the Woods, and, by the 1720s, Lake Winnipeg. It had been largely abandoned by the end of the eighteenth century in favour of a shorter route to the south that travelled up the Pigeon River. But when Jay's Treaty assigned Grand Portage (now in Minnesota) to the United States in 1794, the NWC was forced to return to the Kamin-

3.1 Robert Irvine, *Fort William, an Establishment of the Northwest Company, on Lake Superior*, 1811. (Courtesy of Library and Archives Canada)

istiquia, where in 1802 it began building the fort it later would name Fort William after its chief partner, William McGillivray.

Securing a location on the far shore of Lake Superior was critical. It marked an indispensable halfway point for a continental trade system propelled only by human labour, in a northern climate with a small window of ice-free months for paddling trade goods up from Montreal or furs down from the Athabasca country – upwards of 1,800 or 2,000 kilometres from either direction. One scholar has suggested that more than any other fact, seasonal opportunity was responsible for the shape of the interior fur trade.[3] The NWC could only negotiate around the climactic constraints of the boreal north as long as it had a base of operations in the middle of the continent.

This base represented a substantial commitment by the company, for both profit and morale. As Gabriel Franchère described it in 1814,

"Fort William has really the appearance of a fort, with its palisade fifteen feet high, and that of a pretty village, from the number of edifices it encloses."[4] Every July, partners and traders from Montreal met with their counterparts from the interior (*les hivernants*, those who overwintered in the country) in a huge rendez-vous that spilled out beyond the fort walls. In the Great Hall where they dined hung a copy of David Thompson's "Map of the North-West Territory of the Province of Canada" – a continent read through a commercial network, a visual declaration of empire.[5]

But the continent was not theirs alone. Thomas Douglas, the Earl of Selkirk, resented the fur trade, and specifically the Nor'Westers, for besieging his nascent settlement at Red River (the future Winnipeg), which sat in the middle of crucial river routes west. In August 1816, Douglas occupied Fort William and arrested the NWC partners there. The legal battle that ensued – between Selkirk and the NWC partners, between the Nor'Westers and the HBC – at a time that fur populations were declining in much of the old fur trade territories, exhausted the resources of the NWC, and it merged with the HBC in 1821. Fort William lost its prominent place in the fur trade, over the next few decades the fur trade lost its prominent place in the continental economy, and traffic gradually declined until the post was closed in 1882.

But now the Canadian Pacific Railway – the new corporate power of the Canadian west – acquired the site at the mouth of the Kaministiquia to serve again as an *entrepôt* in another corporate empire and transcontinental system, this time shipping grain from the prairies out through the Great Lakes. The last of the post buildings were demolished by 1902, and the original Fort William lay buried under the CPR rail yards – fifteen sets of tracks – while its namesake city grew up around it. The economic base shifted to pulp and paper, to capitalize on Ontario's boreal forest, and heavy manufacturing. But by the 1960s, lake shipping was undercut by highway trucking, logging and mining faltered, and Thunder Bay saw the same suburban flight as other North American cities. The waterfront was locked into a rusting infrastructure of industrial decline.[6]

In the national story, though, the lakehead remained firmly lodged in the eighteenth century. In 1923, Fort William was among the first places to be designated a national historic site by the Historic Sites

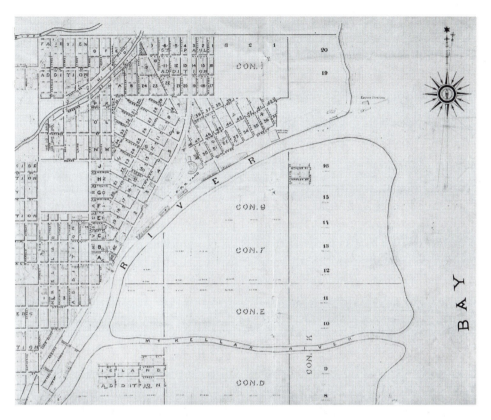

3.2 D.G.W. Richett, *City of Fort William, Ontario,* 1906.
(Courtesy of Library and Archives Canada)

and Monuments Board (created in 1919). This was not surprising, not just because of Fort William's historical prominence, but at least as much because the fur trade and the boreal wilderness were so central to nationalist historiography and popular culture in Canada. The HSMBC's favourite subjects – exploration, continental occupation, the North, French–English conflict, and natural resources – meant many of its earliest choices were drawn from the fur trade: Prince of Wales Fort, on Hudson Bay (1920), Fort William (1923 and again in 1968), Fort Augustus and Fort Edmonton on the North Saskatchewan River (1923), Fort Langley, Fort Kamloops, and Fort Victoria toward the Pacific (1923 and 1924), and so on.[7] Fur trade posts became popular choices for historical reconstruction and reenactment by the middle part of the twentieth century, for precisely the same reason: how they resonated with ideas of Canadian identity in a northern environment. With the largest city in Northern Ontario facing an uncertain economic future, it was only a matter of time before a new Fort William was proposed.

The Fort William Project, 1970–1975: The Value of Historic Reconstruction

The announcement came on 20 January 1971 from Ontario premier John Robarts. Six months later, on 29 July, the project of reconstructing Fort William was contracted to National Heritage Limited, a private research company that had already produced a feasibility study titled *Fort William: Hinge of a Nation*.[8] Initially, the Fort William project was the responsibility of the Department of Tourism and Information, which also oversaw other popular reconstructions such as Sainte-Marie among the Hurons in Midland. In 1972, these were transferred to the new Historical Sites Branch in the Division of Parks, in the Ministry of Natural Resources. The conversation (at times quite combative) between National Heritage, the province, and the historians and archaeologists involved in the project gives us a wonderful picture of the political, pragmatic, and academic issues that play out in a large-scale heritage project.

In the early 1970s, historic reconstructions were hot properties, politically speaking. Living history sites like pioneer villages were proven successes in a postwar North America fixated on family and automobiles. The number of visitors to National Historic Parks tripled in the 1960s, confirming heritage tourism as a growth industry.[9] Whether trading forts or pioneer villages, the big reconstructions all had one thing in common: *where* they were mattered almost as much as *what* they were. They needed to be outside of modernity, but not too far. They needed enough space to be credible as islands of a past era, as "magic kingdoms" enclosed from the outside world; and enough space to host parking lots, and any number of revenue-generating activities. They ought to be located in communities that needed an economic boost, but near enough to prosperous urban markets. The

National Parks Branch was at work on the largest reconstruction in its history at the Fortress of Louisbourg, a project that demonstrates how many competing expectations could be placed on a historic site. It was to recreate one of the most imposing statements of French presence in early Canada, and thereby present a reassuring message of the French *in* Canada during the Centennial era; showcase the emerging talent in public history in Canada; and generate employment as coal mines were closing on Cape Breton.

Louisbourg had federal resources behind it, and the academic expertise of Parks Canada at its height. But the provinces believed they, too, could build their own Louisbourgs. Developing historic sites would allow them to assert more authority in cultural matters, craft distinctive images for themselves within the federation, and exploit new sources of revenue. Alberta, for example, began acquiring fur trade sites in the late 1950s to encourage tourist traffic in remoter areas of the province. Ontario's Historic Sites Branch hoped Fort William would be "precedent-setting" in scale, rivaling the behemoth at Louisbourg.[10] Fort William would seed economic diversification in Ontario's north, a regional counterpart to popular sites on the St Lawrence like Old Fort Henry (a restored fort) and Upper Canada Village (an invented grouping of relocated pioneer buildings), and Sainte-Marie among the Hurons (a reconstructed mission).[11] The commercial viability of public history – its ability to be financially self-sustaining – was one of its more appealing qualities for politicians, but of growing concern to historians. Ontario's provincial archivist, D.F. McOuat, actually thought Fort Albany was the best opportunity for a "very accurate reconstruction" of an eighteenth-century HBC post, but added resignedly that "without road or rail access or acceptable tourist facilities, it could be justified only as an indication of our historical heritage worth doing for non-material reasons alone."[12] The neoliberal climate was, by the 1970s, a significant part of the landscape.

As a destination, Fort William would have two great advantages, and neither was "non-material." One was the size and completeness of the original complex, the "pretty village"; the other was the vibrant colour and action of its signature rendez-vous. In other words, it would be physically impressive as well as exciting to animate: "a bustling and diverse center of commerce and cavorting," a site of business and pleasure, of camaraderie and good food in the middle of a wilderness.[13] National Heritage proposed to reconstruct Fort William to the 1816–20 period. These years following Selkirk's occupation of the fort generated a great deal of documentation, in legal records and post inventories, but it also represented the height of NWC power in the interior, so we "catch the Fort at a time when it afforded the highest degree of interest and variety as a spectacle."[14] *Spectacle*, tellingly, was a word that National Heritage used often.

Whereas the Hudson's Bay Company was based in London, the Nor'Westers could be seen as more Canadian by virtue of being headquartered in Montreal, and embedded in the rivers and lakes of the continental interior. As Alan Gordon has noted, fur trade sites staked out the geography of the Laurentian thesis of Canadian history, a story that celebrated east–west dynamics and economic sovereignty in North America.[15] Fort William represented "the buckle holding together East and West, the vital link, fixing the future path of the transcontinental railroads."[16] Such a territorial version of Canadian history was conservative even in 1971, but it was precisely *because* of its conservatism that it so appealed to post-Centennial Ontario. It neatly suppressed any rival "limited identities" – regional, racial, or otherwise; restored authorship of the national experience to central Canada; and made Ontario's historic sites were much more than provincial in importance. Rivers like the Mattawa and the French, cradled in glacial fault lines and spillways, were referred to as highways of the nation; Fort William as "hinge of a nation" suggested an emergent national capital. The Lake Superior shoreline, the blue lake and rocky shore of campfire song, was thought to be "quite capable of demonstrating the history of Canada in miniature."[17] According to *Hinge of a Nation*, Fort William encapsulated the story of Canada in its continental expanse, its gift of multicultural diversity, and its wilderness identity:

It was Canadian, then, typically Canadian. It was the command post of a great commercial enterprise of a far-flung, intricate organization, served by the skill of French craftsmen, by the experience and daring of French canoe-men, by the woods-lore of the Indian, by the canny, long-range scheming of the Scottish

merchants, and by suppliers and hangers-on of every racial identity ...

Yet they were founding Canada geographically, even as they
were beginning to live our country's characteristic life. Could
people of such different origins, languages and standards make
common cause? The fact is that they did ... It furnishes a salu-
tary object lesson, in miniature, if you like, of what the whole of
our country has been attempting ever since ... No other historical site in our possession, if fully restored, could teach us so
much about ourselves.[18]

On 3 July 1973, Queen Elizabeth and Prince Philip arrived in Thunder Bay to officially open Fort William Historical Park, though only
three buildings were ready (the house of superintendent James Taitt,
the house of independent trader Jean-Marie Boucher outside the fort
proper, and the naval shed). The pomp and circumstance of the royal
visit – which included a ceremonial rendez-vous with demonstrations
of Indigenous dancing, pipe bands, and wood-chopping[19] – camou-
flaged growing discord between National Heritage's directors and its
staff historians, and between the company and the new Historical Sites
Branch. In these first few years, the Branch had become increasingly
concerned with the historical accuracy and the authenticity of the reconstruction. Their complaints about National Heritage were many
and varied, but centred on three principal issues: the depth of historical research, the dismissal of archaeological findings, and the degree to
which profitability drove the agenda. Here I want to focus on the last
of these, as well as the greatest, and yet least discussed, failing of "authenticity": the status of the environment in historical reconstruction.[20]

"Profit Will Carry the Field": Private Initiative in Public History

As a complete manufacture, Fort William Historical Park was more
vulnerable to academic critique than restoration of an existing structure might have been, but the transparent commercial agenda exacer-
bated already short academic tempers. The provincial government was
obviously complicit in this strategy, starting with the initial decision

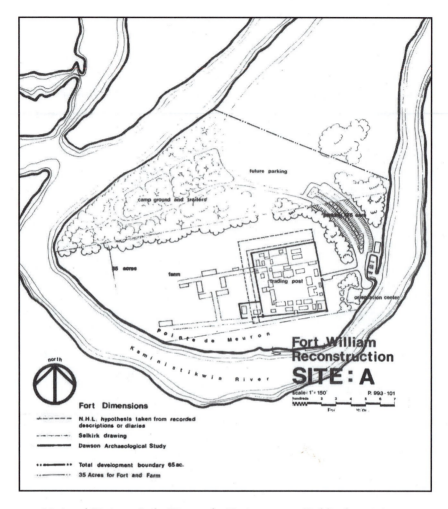

3.3 National Heritage Ltd., *Hinge of a Nation*, 1970. (Public domain)

to rebuild upstream. Preliminary archaeological digs had been underway at the CPR rail yards since 1965.[21] But there was no way that fifty
wooden buildings could be constructed in the middle of downtown
Thunder Bay, and no way that the Ontario government was prepared
to pay market rate for the location on top of the project's budgeted
$10 million. Nor was the city of Thunder Bay interested in using her-
itage as part of its strategy for downtown renewal in the late 1960s.[22]
At Pointe de Meuron, though, the province could afford a full 125

acres with space for a proposed "recreational and tourist centre" surrounding the fort itself. (A similar process was underway in Edmonton; the last original fort in the city centre had been dismantled during the First World War, but was being reconstructed, with no original materials, several kilometres upriver on a site that could provide more spacious parkland).[23]

This meant excavation and reconstruction would proceed in tandem, at entirely different sites. National Heritage argued that a quick reconstruction was essential in order to generate a return as soon as possible. Historians on its staff and with the Historic Sites Branch disagreed. Several researchers at National Heritage Limited resigned or were fired in 1972 after they argued that more research was necessary. That same year, Brian Woolsey complained, "Only after the archaeological and the historic research has been completed and evaluated, should we begin to plan the interpretation of a site. Still less should we decide to reconstruct an historic site in the absence of this information. *As everyone knows*, however, this simple rule was not obeyed at Fort William."[24] The archaeologists at work on the rail yards were aggravated by the reconstruction, which they felt ignored their preliminary findings in order to stay on schedule. "Everyone agrees that research should precede reconstruction," wrote the archaeologist leading the dig to the director of the Historical Sites Branch in 1974. "Everyone acknowledges, but no one accepts the fact, that, in the case of Fort William, it has been done simultaneously ... You can't have the answer, because it doesn't exist yet."[25] The push for completion pressured archaeologists at the Pointe de Meuron site as well. A frustrated Donald MacLeod let loose over what he saw as research sacrificed to deadlines: "Ethnohistoric research and follow-up *excavation on Indian camps* across the river is *essential*. I think this should take priority over European architecture, in terms of *interpretative planning* – if we were not geared to a @* construction – *deadline and the profit motive!*"[26]

But businesses run on such deadlines, and both National Heritage and the province wanted a complete reconstruction, not a complete excavation, as soon as possible. Once the doors were open the new fort could generate revenue through admission fees, as well as by keeping people in the area as long as possible and by making sure they bought something. The first was a familiar strategy for regional development; the second was particularly well suited to a fur *trade* post. The fort would be encircled by a campground, possibly even a marina, to encourage visitors to remain for a few days. The park store would sell gas and camping supplies. The pack store, Indian shop, dry goods store, and liquor store would be used as "spectacle, entertainment, or shopping" venues.[27] This tourist traffic would generate income for local businesses, and "careful authentic reproduction of the antiques and artifacts proper to Fort William" – everything from candlesticks to snowshoes, and "of course a considerable stock of hand-made Indian productions" – would provide quality souvenirs.[28]

By 1972, the Branch was becoming increasingly forthright – and occasionally vitriolic – in its opposition to the commercial proposals for the site. The promotion of Fort William was likened, disparagingly, to that of Disneyland.[29] When National Heritage suggested draining the adjacent marshland to create a more scenic waterway, Woolsey insisted it was "artificial, unhistoric and ecologically perilous."[30] In 1972, he confided to Fred Armstrong at the University of Western Ontario, "The major flaw in the whole affair, in my opinion, was allowing a profit-oriented organization to plan and reconstruct a historical site ... In a crunch, profit will carry the field over authenticity ... Like you and Dr. Careless, my colleagues in the Division of Parks and myself believe that there is more to a historic site than tourism, restaurants and profits. We shall do our best to make that view prevail."[31] Barely begun, Fort William was *already* seen as a cautionary tale. In 1971, the Branch opted to undertake its own archaeology at Michipicoten Mission to "protect our investment" in research.[32] Even more revealing was the Branch's reaction to a 1972 proposal by a private firm to develop a "historic" site about mining in Cobalt. Woolsey's scathing disdain for the profit imperative, and the importance of professional and political reputation, stemmed directly from his experience with Fort William:

> The frequent mention of engineers in connection with the restoration [at Cobalt] would seem to indicate that there will be no more than lip-service paid to historical authenticity ... What is

most alarming about this planning document, however, is its unblushing concern for tourist dollars at the expense of visitor experience. Almost every time an historical building is mentioned, it is in connection with a café, gas station, or some other irrelevant tourist outlet. [It] is a poor effort, which no professional historian or architectural historian would be associated with if he valued his professional credentials. If this preliminary plan is accepted by Industry and Tourism, the Ministry of Natural Resources should not become associated with it. The planning (or lack thereof) for Cobalt *has all the markings of another Fort William.* When will people realize that the best guarantee of tourism is a professional high quality historical product?[33]

Today, Fort William sells itself as both historic and usable space. It offers banquet facilities with "fur trade décor ... complemented by an inviting view of the surrounding natural environment" as well as a "pristine wilderness setting ... for RVers and campers."[34] Its calendar of events, however, has little to do with either the fur trade era or the history of the area; recent years have seen performances of Shakespeare's *Taming of the Shrew*, Charles Dickens's *A Christmas Carol*, Canadian rock musicians, and the Canadian PGA tour, in addition to reenactments of the summer rendez-vous or the "Battle of Fort William" of 1816. The site can be rented not just for school groups but for private functions such as weddings and conferences. But we have seen the same evolution in the national sites system, particularly since the creation of the Parks Canada Agency in 1998. And this is key. After all, the commercial impetus makes far more sense at Fort William than, say, Louisbourg. After all, Fort William was a *trading post*, a place that anchored imperial commerce and exhaustive resource harvest. That the reconstruction was contracted to a private company is thus more apt than terribly surprising. What is more revealing is the role of the provincial government. For all the protestations of the historians, the Ontario government's decision to reconstruct a trading post, as part of a strategy of economic development, sanctioned the continuing story of Canada as a corporate enterprise.[35]

Finding – or Making – a Fur Trade Wilderness

When scholars have talked about the authenticity of historic sites, they have usually framed their critique much like Woolsey did: commerce as opposed to education, research versus image. What we haven't considered is the environmental meaning, value, and cost of these sites. What did it mean for both the authenticity and profitability of the reconstructed Fort William to be located in a different place than the original? And what does the choice of site say about our perception and expectation of nature in Canadian history?

A historic site wants us to adopt the same perspective as those who came before, to believe we are standing in their footsteps. This lends credibility to the site and makes the past seem more accessible, less abstract. We can't begin to understand why people acted as they did, where they did, until we have at least an imaginative approximation of the world they saw around them. In this, Pointe de Meuron was a better facsimile of the eighteenth-century Kaministiquia than the rail yards. Its partial isolation offered a spatial buffer to minimize the intrusion of modern sightlines and to nurture a suspension of modern disbelief.[36] A fur trade post – and especially one that hinged a transcontinental trade system – needed to at least suggest the boundless interior and spatial ambitions of the eighteenth century. A winding trail from the parking lot to the palisade would "help immerse [visitors] in nature and realize the natural elements which influenced history."[37] In reconstructing Fort William, immediate *atmospheric* authenticity took precedence over broader *locational* authenticity.

Accessing "our country's characteristic life" required accessible fragments of such an environment. Popular Canadiana, especially as conceived by central Canada in the postwar period, drew heavily from the artifacts and geographies of the fur trade. It was bicultural and transprovincial, heroic and relatable, and expansive in a way that mapped contemporary Canadian boundaries.[38] When discussions began about reconstructing Fort William, Canada was still on a high from the 1967 centennial of Confederation, the celebration of which included a voyageur canoe race along the old NWC route from Rocky Mountain House to Montreal. The American National Parks Service noted a surge in visitation in 1967 at Grand Portage National Historic Monument –

the site of the NWC entrepôt before its relocation to the Kaministiquia – by Canadians retracing the route of the voyageurs in the Centennial year.[39] Reconstructing Fort William would help reclaim some of that tourist traffic from America, and back to its rightful home!

So this investment was as much in the northland as in the fur trade. "Tourists from far and near" had arrived at Fort William by the 1870s, attracted by both the romance of the Nor'Wester fort in ruin and the sport fishing.[40] It was very close to the height of land that divided water flowing south to the Great Lakes from that draining north to Hudson Bay. This divide has had a particularly mythic quality in a country seeking a northward identity despite its southward realities:

> To the last portage and the height of land – :
> Upon one hand
> The lonely north enlaced with lakes and streams,
> And the enormous targe of Hudson Bay,
> Glimmering all night
> In the cold arctic light;
> On the other hand
> The crowded southern land
> With all the welter of the lives of men.[41]

That the north most would see was only a few hours' travel from Toronto – the Muskoka lakes, Algonquin Park, Georgian Bay – was immaterial; crossing onto the Canadian Shield from the St Lawrence lowlands was evidence enough of a northern character. "To all of us here, the vast unknown country of the North," said Stephen Leacock, writing in Orillia, "supplies a peculiar mental background."[42]

By the 1960s, those men (and women) from the crowded south were recolonizing the North in record numbers, intent on becoming modern-day voyageurs by canoeing the lakes and rivers of the Shield. Pointe de Meuron represented a sizeable green space for the city of Thunder Bay, but was also an important asset for the province, whose Parks Branch was responsible for natural resources as well as historical ones. Demand for canoe routes in Killarney and Quetico – both along the old NWC routes, and the latter only 160 kilometres west of Fort William – led to their designation as the first wilderness provincial parks in 1971. Following fur trade routes was a means of reconnecting with a national tradition on an individual level. Consider the language used to promote the Mattawa, a key part of the NWC route from Montreal, designated the province's first "wild river" park: "For the modern voyageur, a trip down the Mattawa is a chance to relive one of Canada's most colourful periods, the fur trade era. As he paddles his canoe downriver, he can, if he tries hard enough, still hear snatches of voyageur song, smell voyageur campfires, and perhaps even shiver with some the same fear of a mysterious, spirit-filled wilderness … [and] understand the difficult task facing the fur traders who took thirty-eight foot canoes and their huge cargo over these same paths."[43] Historian W.L. Morton had written that the "alternate penetration of the wilderness and return to civilization is the basic rhythm of Canadian life," a rhythm that could describe voyageurs and weekend canoeists alike.[44] Poet Douglas LePan described canoeists launched into a "time without tense" where through the mist appears a ghostly ancestry, "paddles flashing / as a brigade of *canots du nord* drives upstream … Two hundred years are nothing, nothing."[45] And in 1968 the country had just elected, in unprecedented popularity, a young prime minister who was a passionate advocate for the national education of the canoe. As Pierre Trudeau wrote,

> What sets a canoeing expedition apart is that it purifies you more rapidly and inescapably than any other. Travel a thousand miles by train and you are a brute; pedal five hundred on a bicycle and you remain basically a bourgeois; paddle a hundred in a canoe and you are already a child of nature.
>
> I know a man whose school could never teach him patriotism, but who acquired that virtue when he felt in his bones the vastness of his land, and the greatness of those who founded it.[46]

The story of Canada – a political improbability, a country that had always to defy geographical and cultural diversity – depended on a mythic and unifying identification with the land, and specifically, with its largest component ecosystem: the boreal forest on the Canadian Shield. The Group of Seven had created one version of this story in the 1920s; the postwar popularization of the fur trade supplied another.

A history *in* this kind of landscape had given us a sense of territory and character – two essential elements of nationhood, and priceless for a country that, in the 1970s, was as anxious as it would ever be over disunity between east and west, English and French, on the one hand, and its ability to exist as the northern part of North America, as not-American, on the other. As Pierre Berton wrote to "Sam," an imagined American correspondent, in 1982,

> An endless expanse of gnarled grey rock, pocketed by millions of gunmetal lakes, with twisted pines, skeletal birches and stunted black spruce bending before the wind. No covered wagon could cross it, only strong men sturdy enough to hoist a canoe on their backs or to shoulder a hundred-pound pack at the end of a tumpline. We are a Shield people, Sam, a wilderness people ...
>
> To the west, the mouth of the long water highway that leads to the heart of the continent and has, as its continuation, the lakes and the horizontal rivers that lead across the prairies and trickle through the mountain passes to give Canada its shape and its reason for being. For we are a nation of canoeists, Sam, and have been since the earliest days ... When somebody asks you how Canada exists as a horizontal country with its plains and mountains running vertically, tell him about the paddlers, Sam.[47]

All this says much more about latter twentieth-century perceptions of – really, yearnings for – an imagined fur trade environment. While hundreds of posts were located on rivers, they were generally situated to take advantage some kind of intersection: a fork, a junction, a portage, or where a river flowed into the sea or a lake. The original Fort William was located at a delta, on low, swampy ground where the Kaministiquia forked around two large islands. This would have affected both the fort's view and capacity. As Franchère reported in 1814, "No site appeared more advantageous than the present for the purposes intended; the river is deep, of easy access, and offers a safe harbor for shipping."[48] One of the first buildings reconstructed at Pointe de Meuron was the naval shed, where the NWC built and re-paired canoes and bateaux, thereby acknowledging (paradoxically) the importance of the harbour location to the original fort. This was

underscored seventy years later by none other than William Van
Horne of the Canadian Pacific Railway, who fought vociferously for
the right to develop the original lakeside site into a complex of rail
yards and elevators to move grain onto lake freighters. As he wrote in
1886, "The Kaministiquia has advantages for coarse freight business
unequalled in the whole of the Great Lakes."[49] The forest screening at
Pointe au Meuron appealed more to twentieth-century tastes for recreation than nineteenth-century ambitions for industry.

And it is important to remember that the fur trade *was* an industry, in spirit and practice. Like most of settler society in the eighteenth
century, fur traders actively "improved" their surroundings. Unlike
the popular image of canoes passing through untouched forest, fur
trade posts sat in clearings: the wood had been felled to construct post
buildings, and as at Fort William, cleared land was required to grow
food. Franchère noted approvingly that "It is true they had to contend
with all the difficulties consequent on a low and swampy soil; but by
incredible labor and perseverance they succeeded in draining the
marshes and reducing the loose and yielding soil to solidity ... The
land behind the fort and on both sides of it, is cleared and under
tillage. We saw barley, peas, and oats, which had a very fine appearance."[50] Food provisioning was a major concern, especially for the
Nor'Westers, who relied on literal manpower for moving supplies in
a pre-petroleum age. Fort William was an important supplier of foodstuffs in farm products and fish. Voyageurs were required by contract
to devote several days a year to clearing land, and many retired nearby
to farm.[51] In one of the great ironies of the reconstruction, the woods
that suggested a boreal forest at Pointe de Meuron were themselves a
recent addition. Seventy years before it was a "bright home-like spot"
with "farm-buildings and a field of ripening wheat" that had been
under cultivation for most of the nineteenth century – since the days
of Fort William.[52] (But it was on the river's flood plain, and the reconstruction has been flooded as a result. In that, it does recall the low,
swampy ground of the original lakefront site!)[53]

When the Historical Sites Branch formally evaluated the Fort
William project in 1974, before responsibility for its operations was
to be transferred to the province, it was appalled. "Nary a building
has gone up to date for which we have not had reservations," wrote

architectural historian Denis Mahon. Invoking a review of the project's authenticity would be "our last stab at rendering Fort William buildings authentic (*forgetting the siting, of course*)."[54] The review was unprecedented in historic sites practice, but Mahon's aside was an admission of its limitations. The choice of site had been a necessary evil; the architecture was the work of human hands and thus correctable. Within a few weeks an Authenticity Committee had compiled a list of *one hundred* points of concern, mostly errors of period detail in material and construction: the use of twine and string instead of hemp or linen cordage, cement in the chinking and plaster, anachronistic furnishings and paint schemes. The committee also compiled a chronology of past complaints dating back to November 1971, most of which had to do with National Heritage failing to submit evidence in support of different building plans, and summing up the problems of the Fort in one damning line: "DETAIL all shot to hell." But it also raised concerns about layout (agreeing with the archeologists on this) and the landscaping; anachronistic elements like neat graveled paths were judged "very inaccurate." The Authenticity Committee stated firmly, "The authenticity of the reconstruction depends on the total environment, not just on the structures and the furnishings."[55] But it could offer few concrete suggestions. Palisades, canoes, blankets, beads, guns – these were well-documented in post records, widely recognized, and accordingly, hallmarks of most eighteenth- and nineteenth-century historic sites. They could be reproduced based on archival research, and thus satisfy both professional and public expectations of a fur trade site. But inventories so helpful in itemizing stores and trade goods offered "no direction" on landscaping at Fort William.[56] Moreover, a palisade or a Great Hall, once built, would stay helpfully in place; any landscaping around it would make no such promise.

Dynamic Wilderness, Past and Present

Such is the paradox facing many public history sites. Structures can be restored, reconstructed, and maintained to a fixed past state; their surrounding ecosystems cannot. Both the site of the reconstruction and the original Fort William had changed dramatically over the course of the twentieth century, largely due to human preferences and interventions. Neither had been a "pristine wilderness" for two centuries. The proxy at Pointe de Meuron was farmland regrown into suburban parkland. Fort William in its heyday had been an important base for farming with lands "cleared and under tillage," and supplying the brigades with food. It was buried under rail yards *because* it was on a lakefront, and therefore useful for moving goods in bulk, first by canoe and then by freighter. But characterizing Fort William as a proto-industrial enterprise did not sit well with ideas of wilderness that had become central to Canadian identity.

Here again, we see that historic sites are not islands separate from their larger environments. The same shoreline locations that made many sites successful as trading posts have made them vulnerable to erosion. Much of the Rocky Mountain House historic site is sliding into the North Saskatchewan River; Michipicoten Post is now cut off from its host river by sand and glacial till eroded from the river bank; York Factory is eroding into the Hayes River at a rate of up to three metres every five years, with at least two full post sites now sitting on the riverbed, while climate change weakens the permafrost on which the posts were built. Archaeologists in the late 1960s noted ongoing erosion on McKellar Island, where voyageur huts had been built across from Fort William.[57] As we saw in Chapter 1, Parks Canada has only just begun to acknowledge the implications of climate change at a few historic sites. More tellingly, we have *not* acknowledged the extent to which these "natural" processes have been accelerated by human activity. Trekking up and down a riverbank is a sure way to hasten the erosion of that very bank.

But Fort William was not designed to convey such change. Just the opposite: it assumed a fixed and imagined wilderness in the public mind and, like most living history sites, it focused on the built environment, to be perfected within the "total environment." Despite – or perhaps because of – the public debates over authenticity, it succeeded. In 1984, a representative from the American Association of Museums judged Old Fort William to be "one of the best examples of Living History" he had seen.[58] It is one of, if not the, most complete representations of fur trade life in Canada. Fort William proved to be a successful reconstruction in terms of provincial strategy and public experience. Contracting the project to a private company resulted in an intense discussion about the meaning and measures of authenticity. But that discussion never really considered the environmental past, how to address environmental change, or most important of all, the expectations of wilderness in the national imagination. Wilderness in Canadian public history would remain an article of faith.

CHAPTER 4

Variety, Heritage, Adventure, and Park: The Forks of the Red River

Canada is a country of rivers. Our national motto – *a mari usque ad mare*, "from sea to sea" – was inspired by Psalm 72: "He shall have dominion also from sea to sea, and from the river unto the ends of the earth." The St Lawrence, the Saskatchewan, the Mackenzie: they have been routes of exploration, means of claiming a vast continent. Even in a road-riven Canada, rivers remain the geographical scaffolding in the national imagination. In his haunting "Northwest Passage," folk singer Stan Rogers positions himself as a modern-day "explorer, driving hard across the plains … to race the roaring Fraser to the sea." In 2004, a member of Parliament from Saskatchewan proposed the national motto be changed to "A nation of rivers, a river of nations," but Bill C-529 stalled after its first reading in Parliament.[1] What does this say? Do Canadians, especially urban Canadians, recognize the historical and ecological significance of rivers around them?

This chapter examines a city often thought of as a Prairie city, the gateway to the golden west: Winnipeg. Like the original Fort William, the Forks at Winnipeg was a fur trade site buried for much of the twentieth century under rail yards. But whereas Fort William was reconstructed upriver, the Forks were unearthed from railway infrastructure in order to provide the city with park space and a signature attraction. It was a massive project of urban reclamation, to reinvigorate a derelict industrial core into an economically self-sustaining complex of farmers' markets and cafés, performance spaces and

4.1 Peter Rindisbacher, *Winter Fishing on the Ice of the Assynoibain & Red River*, 1821. (Courtesy of Library and Archives Canada)

public sculpture, and riverside park. The Forks' historical designation as a "meeting place" sanctioned both commerce and recreation.

The result is a success in many ways. The Forks National Historic Site is Manitoba's most visited attraction, with more than four million visitors each year. It is a civic symbol of Winnipeg, with restored railway buildings alongside postmodern architectural designs, riverview pathways through patches of urban forest. It is a "meeting" of different agendas, different uses, and different kinds of landscape. However, this multiplicity of purposes has required the Parks Canada property to be more urban park than historic site, with a relatively muted historical presence. Originally, though, the Forks were imagined in a larger context of a regional river park, and a clear call for environmental citizenship. Could we take inspiration from these earlier plans to enliven the environmental lessons of the Forks?

A Meeting Place

The Forks has been referred to as "the single most historic site in all of Western Canada."[2] It's a fair assertion. For six thousand years, people have travelled to the junction (the "forks") of the Red and Assiniboine Rivers for trade and sustenance. The river valleys formed a kind of transition zone between the northerly parkland and the western plains, rich in resources and easily accessible. Nakoda (Assiniboin), Cree, Dakota, and Anishinaabe met here to trade, fish, and hunt, part of a dynamic and diverse landscape of peoples in competition, warfare, and exchange that extended across the continental interior.[3] After Sieur de la Vérendrye established Fort Rouge here in 1738, rival fur trade companies erected a series of posts (Fort Rouge, Fort Gibraltar, Fort Douglas, and Fort Garry) on opposite shores of the rivers to tap into these aboriginal trading networks. The rivalry between the Hudson's Bay Company (HBC) and the North West Company (NWC) intensified especially after Thomas Douglas, Earl of Selkirk, landed settlers here in 1812, inciting new conflicts with the growing population of Métis, who had emerged as a crucial force in the buffalo hunt. They were joined soon after by a new French-Canadian presence with the founding of St Boniface on the eastern side of the Red River in 1818. After the amalgamation of the two companies in 1820, the HBC governed its vast holdings of Rupert's Land from Upper Fort Garry at the Forks here, and when turning over the territory to the new government of Canada, the Forks and its fort became the symbolic centre of Métis nation led by Louis Riel.

With the decline of the southern fur trade in the mid-nineteenth century, the HBC began reusing the river flats in ways that typified the new interest in the prairie west: assigning river-lot farmsteads, establishing an experimental farm, building grist and flour mills.[4] The remaking of the prairie from grassland to grainland followed the new rail lines running across the continent, and they ran through the Forks. (Upper Fort Garry was levelled in 1882 to permit the construction of Main Street directly through it.) In 1888, the Hudson's Bay Company's Reserve – a 465-acre parcel of land at the forks – was bought by the Northern Pacific and Manitoba Railway, then leased to the Canadian Northern Railway and its successor, the Canadian

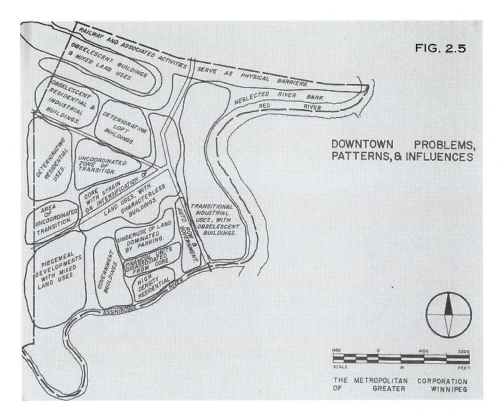

FIG. 2.5

RAILWAY AND ASSOCIATED ACTIVITIES SERVE AS PHYSICAL BARRIERS.

OBSELESCENT BUILDINGS & MIXED LAND USES

OBSELESCENT RESIDENTIAL & INDUSTRIAL BUILDINGS.

DETERIORATING LOFT BUILDINGS.

NEGLECTED RIVER BANK.
RED RIVER

DETERIORATING RESIDENTIAL USES.

UNCOORDINATED ZONE OF TRANSITION.

CORE STRAIN WITH INTENSIFICATION OF LAND USES, WITH CHARACTERLESS BUILDINGS.

AREA OF UNCOORDINATED TRANSITION.

TRANSITIONAL INDUSTRIAL USES, WITH OBSELESCENT BUILDINGS.

UNDERUSE OF LAND DOMINATED BY PARKING.

AUTO ROW & GOVERNMENT

PIECEMEAL DEVELOPMENTS WITH MIXED LAND USES.

GOVERNMENT BUILDINGS

FINANCIAL UNITS DISASSOCIATED FROM CORE.

HIGH DENSITY RESIDENTIAL

RIVER

ASSINIBOINE

DOWNTOWN PROBLEMS, PATTERNS, & INFLUENCES

1600   0   1600   3200
SCALE   IN   FEET

THE   METROPOLITAN   CORPORATION
OF   GREATER   WINNIPEG

4.2 Reid, Crowther & Partners, *A Market Analysis of Metropolitan Winnipeg*, 1967. (Public domain)

National. Immigrant sheds housing those bound for the advertised golden prairie gave way to a purely industrial zone of railway fill, storage, and fuel depots cut off from the city by berms and embankments. Meanwhile, these immigrants and the railway cars that carried them built a grain empire symbolized by the grand "skyscrapers" in the Exchange District, as the commercial centre of Winnipeg – a self-proclaimed Chicago of the North – moved away from the Forks.[5] By the 1960s, CN was shifting its operations out of the downtown to cheaper land east of St Boniface. Winnipeg, like many other North American cities, was increasingly concerned about the economic (if not ecological) health and visual unsightliness of this "drab section of our core." Municipal reports cited "the shopworn fabric of a downtown riddled by decades of neglect," obsolescent infrastructure blocking public access to a neglected riverbank.[6] Industrial use had interrupted millennia of meeting.

As with Fort William, concerns for urban renewal here transected beautifully with well-established narratives of Canadian history. From its inception in 1919, the Historic Sites and Monuments Board of Canada favoured sites that emphasized heroic stories of nation building like continental exploration and settlement of the interior. Canadian history was dominated by writers such as Donald Creighton who saw a national destiny unfolding along continental rivers. Positioned on the edge of the prairie, hinting at the vast territories that lay beyond (and which pre-dated provincial boundaries), the Forks reaffirmed a nationalist transcontinental sensibility. Its heritage resources – railways, bridges, and landmarks like the 1911 Beaux-Arts Union Station – were all themed to transportation and movement, a crossroads of Canada.[7] Designation as a national historic site in 1974 strategically preserved a federal presence in the heart of a Prairie province amid rising western regionalism. "Why should I sell your wheat?" Prime Minister Pierre Trudeau asked Manitoba farmers in 1969. It was a rhetorical question that, unfortunately, seemed bizarrely insensitive to deep feelings of alienation among westerners who saw Ottawa as both distant and exploitative.

In this context, the Forks could mean something distinct and distinctly valuable to westerners. After all, Louis Riel – by the grace of God the father of Manitoba – is buried just across the river at the St Boniface cathedral. While all the provinces pursued heritage projects in the postwar years, during the 1970s Manitoba was especially active in developing a series of historic sites along the Red River that commemorated overtly regional stories: the Riel family home at St Vital; La Barrière, where the Métis challenged Canadian surveyors in 1869; St Norbert, as a marker of French-Canadian migration to establish a francophone and Catholic community. Presenting the Forks as "the birthplace of Western Canada" could mean either an expansion of national territory or the creation of a new region, depending on if one put the emphasis on "western" or "Canada."[8] In fact, this never became a serious point of contention. The Forks redevelopment required commitments from all three levels of government, but more importantly, it could be easily segmented by jurisdiction and by use.

Redeveloping a River

From the outset, the Forks was to serve two functions: as an entry to a river park, and as an urban attraction. To park planners, these were not contradictory – indeed, just the opposite. In 1972, Prime Minister Trudeau announced a new Byways and Special Places Program, to adapt historic river corridors for regional economic renewal, especially recreation (after all, this was the prime minister who had proclaimed that patriotism was to be learned by paddling). The next year Ottawa announced another program, the Agreement for Recreation and Conservation (ARC), in which provinces could secure funds for redeveloping sites with historical significance and recreational opportunities especially for urban Canadians.[9]

Through the 1970s, numerous studies noted the particular assets of the Red River: prairie riverine ecosystems, historical resources of national significance, provincial support, and most of all, accessibility to a growing urban audience in need of parkland, ideally on foot. This bundling of recreation and conservation was prescient, because it assumed and required a coexistence of but division between commercial and park space. A 1976 report from Parks Canada suggested that "Open parkland at The Forks could be owned and maintained by the federal government in association with the historical interpretation program," boat docking could be managed by both the federal and provincial governments, and "private concessions" could be allocated separately.[10] In 1978, Ottawa and the province of Manitoba signed the Canada–Manitoba Agreement for Recreation and Conservation of the Red River corridor, which awarded $13 million for developing a linear park running from the La Salle River south of Winnipeg through to Netley Marsh at Lake Winnipeg. The Forks, "in the heart of Winnipeg," would be the focal point and the gateway for the entire project. Years before Parks Canada announced Canada's first national urban park in the Rouge River Valley in Toronto, it was in fact describing the Red River and the Forks as such.[11]

As an urban river, however, the Red was not yet in much condition to be a park. There was special concern with the stability of the riverbank. Manitobans had been worried about the Red's flooding for over

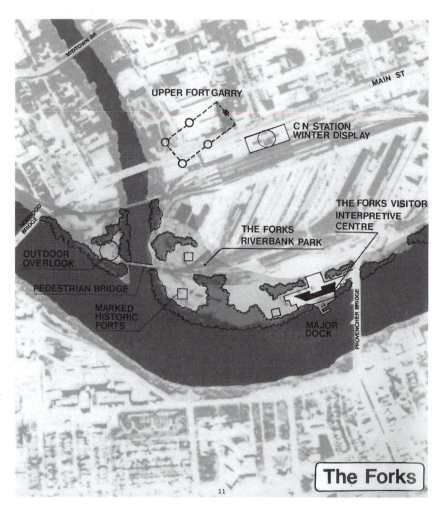

MIDTOWN BR

UPPER FORT GARRY

MAIN ST

C N STATION
WINTER DISPLAY

THE FORKS VISITOR
INTERPRETIVE
CENTRE

THE FORKS
RIVERBANK PARK

NORWOOD
BRIDGE

OUTDOOR
OVERLOOK

PEDESTRIAN BRIDGE

MARKED
HISTORIC
FORTS

MAJOR
DOCK

PROVENCHER BRIDGE

11

The Forks

4.3 ARC Management Board, *Red River Corridor Master Development Plan*,
1981. (Public domain)

a century, and postwar subdivisions intensified pressure and populations on the floodplain south of Winnipeg in particular. At the Forks
proper it was difficult to tell where the industrial fill left off and the
river began: the flats had been filled with a century of cinder, rubble,
gravel, "disposed railway paraphernalia" and other waste that in
some places extended nearly twenty feet deep and had significantly
changed the slope of the riverbanks. In addition, the site would have
to be reconnected with the city from behind its barriers of rail and
road lines, with boat services and walkways. At this stage, however,

environmental concerns were largely aesthetic. References to the river tended to emphasize the importance of vistas or sightlines that would appeal to visitors.[12]

The Red's face, after all, was its fortune. It needed to appeal to large numbers of people, but not simply as green space; it needed to generate money for the city. As site planning accelerated in the mid-1980s, evaluating market potential and user numbers became a major focus. As one historian has noted, urban waterfronts have become attractive sites for public history, but municipal officials prefer the more remote (and generally less contentious) preindustrial past, and consumer-oriented projects.[13] Parks Canada had recently completed such a showcase project in Halifax, with the rehabilitation of seven nineteenth-century warehouses on Lower Water Street. Historic Properties offered ambiance, harbour views, downtown renewal, and commercial space – an ideal combination to attract tourists.[14] While Ottawa and the province would not agree on everything, they could certainly agree on its potential contribution to tourism and Winnipeg's economic health. The initial public funding from ARC and other programs was to jumpstart the engine of economic renewal, but the Forks were to be "always working toward an ultimate goal of self-sufficiency."[15] In 1987, the three levels of government established the Forks Renewal Corporation (FRC) to oversee the redevelopment. The auxiliary train tracks were pulled up by 1988, leaving a number of warehouse structures as well as a century-old bridge crossing the Assiniboine River.[16]

By now it was clear that there were really two Forks and two authorities at the site. Parks Canada would supervise a "heritage park" of 3.6 hectares with primarily archaeological features, while the FRC would oversee the redevelopment and tenancy of a 22 hectare-parcel of former rail grounds. Initially, it entertained proposals for nearly anything with "year-round potential," from a theme park to an aquarium, although a farmers' market and a provincial tourist centre emerged early as the most likely candidates.[17] The Forks Market was created by bridging two stables (1909) with a glass canopy to form a consumer-friendly interior; the adjacent Johnston Terminal (built 1928) was converted to more shops and restaurants. Over the next decade, the Manitoba Children's Museum moved into the Northern Pacific and Manitoba Railway Company's Buildings and Bridges Buildings

(1889), and the CN steam plant (1947) became home to a television studio. With this first generation of occupants housed in the cluster of former railway structures, the site retained an important degree of visual and structural cohesion.

But already there was concern about the market agenda prevailing on historic land. Heritage Winnipeg, the Manitoba Heritage Committee, and the provincial Ministry of Culture, Heritage and Recreation repeatedly complained that historical content was being sidelined in favour of commercial development. The *Winnipeg Free Press* observed pointedly, "Most of the history seems likely to be lost in the shuffle," and a few months later, cautioned, "in the effort to meet commercial viability there will be a grab-bag of projects."[18] The commercial imperative seemed to intensify after 1995, as government funding withdrew. The FRC and its successor, the Forks North Portage Partnership, found itself facing public forums at which Winnipeggers said bluntly, "we were asking for more park, not more parking."[19] But though the FRC was intent on courting corporate sponsors, it did not pursue the condominium strategy found in other historic cores, such as Griffintown and along the Lachine Canal National Historic Site in Montreal. Though they supply usefully dense urban housing, such projects limit public access spatially and through inflated property values. As *Canadian Architect* described the Forks in 2004, "Make no mistake, the site operates as a business to develop the land and invest in public amenities."[20] The key word, though, is *public*. The focus on tourism, which positioned the Forks as an "attraction," and the original designation of "meeting place" reiterated often enough in site planning documents, kept it out of the real estate market. But as the critics noted, the various projects seemed to take advantage of historic space without conveying much, if any, history.[21]

Ironically, it was the site's very historical identity that encouraged, or at least forgave, this new eclectic and ahistorical character. By characterizing the Forks as a meeting place, whether six thousand years ago, two hundred years ago, or last Saturday, we can insert present-day activity into a genealogy of use. But this elasticity has been tested by a series of increasingly ambitious, self-consciously modern architectures: the CanWest Global Performing Arts Centre (1999), the Inn at the Forks (2004), a skate park (2006), and a proposed (but quashed) water park (2013). Now these have been dwarfed (literally) by the Canadian Museum for Human Rights, which opened in 2014 north of the Esplanade Riel footbridge. The architect's design statement promised a building inspired by archaeological memory and prairie ecology at the site, "a symbolic apparition of ice, clouds and stone set in a field of sweet grass. Carved into the earth and dissolving into the sky on the Winnipeg horizon."[22] Poetics aside, this "apparition" is a very corporeal entity, with twelve floors incorporating thousands of tons of imported materials, and a hundred-metre (328 feet) tall tower. Not surprisingly, Parks Canada expressed significant concern over the impact on its adjacent property, in archaeological disturbance, shadows, and sightlines.[23]

Parks Canada and Park Space

This is the context in which Parks Canada has had to negotiate in administering a small parcel of park/land within a larger public/commercial complex. This context has defined a particular role for The Forks National Historic Site: as green space to complement the commercial and cultural activities, so that the Forks as a whole offers a balanced urban playground. Initial plans in the mid-1970s had a major interpretative centre located on the Parks section, or even a reconstruction of Upper Fort Garry. But Parks Canada shelved the idea of a visitor centre in favour of interpretation based on seasonal events, reenactments, pageants, and a nebulous notion of "place." A green space "including such things as theme-related play structures and site furniture (benches, picnic tables, water fountains)" sounds much more like a conventional city park than a historic site.[24] What happened?

In the two decades since Old Fort William rose anew on the Kaministiquia, there had been a noticeable retreat from full reconstructions in favour of ambiance. Public history and public coffers preferred open space and the suggestion of historic (often archaeological resources) to convey a "sense of place." At Melanson Settlement in Nova Scotia, visitors were to gaze over the grassy slopes of the Annapolis River and think about the vanished Acadian homestead beneath; at Batoche they could walk through old Métis river lots down to the battle site on the North Saskatchewan. This approach was more expansive than the earlier practices of a plaque, restoration, or reconstruction, and particularly suited complex landscapes, rural site, and displaced communities. There are also pedagogical arguments for a lighter footprint. Unlike a built restoration, it does not commit the site to one period of occupation; it allows for discussing historical periods that have not left physical remains; and it draws attention to the setting. Consider Rocky Mountain House National Historic Site in Alberta, developed in the late 1970s, which consisted of four separate fort sites on an eroding riverbank. Unlike older fur trade sites such as Lower Fort Garry or Fort Langley, with the rebuilt complexes and costumed interpretation that signalled a federal project of the 1950s, Parks Canada presented Rocky Mountain House as a historic park. It experimented with different kinds of interpretation at each fort location, ranging from an open metal structure that mimics the outline of a fur trade post to interpretative panels at an overgrown archaeological dig site. A small play fort next to the museum is the most complete "fort" on site. Interpretation that stops short of committing to full reconstruction or re-enactments also permits site operation even if research is tentative, or if money is short. "With all the likely sources of historical information exhausted and with a good deal of archaeological research completed, one is left with a set of very inconclusive conclusions," concluded the archaeologist at Rocky Mountain House in 1976, and the site still opened to the public three years later.[25]

The Forks looked far more like its compatriot in Alberta than it did Fort William. Since the primary goal at the Forks was to make use of a downtown parcel of land, a relocated reconstruction in a more "fur-trade-appropriate" setting was a non-starter. The National Historic Site opened to the public in 1989, after years of archaeological digs that continued in the form of popular and precedent-setting public archaeology programs.[26] The permanent installations were minimal: text panels, an "orientation node," a boat dock, a wooden play structure for children, and a riverside amphitheatre etched with pictographs depicting a prairie fire. Pathways with sculptures and views to St Boniface wound through the open space and a small urban forest of dogwood, Manitoba maple, and cottonwood. The most prominent sculpture was located in the orientation node; *The Path of Time* by Marcel Gosselin (1991) features a sundial that silhouettes a parade of past occupants as the sun moves across its face. There was a not a reconstruction in sight apart from a York boat in the children's playground.

Interpretation consisted of walking tours, and a calendar of festivals and events frequently unrelated to the history of the Forks itself. Here again the designation of meeting place proved convenient, because it seemed to justify a rather more spontaneous, event-driven approach. But as one consultant observed in 1989, by emphasizing "sense of place" instead of specific features, the HSMBC and Parks Canada had created a comparatively abstract historic site.[27] Park planners tried to historicize the event calendar by suggesting everything from historic garden plots from different ethnic traditions to cariole races and tobogganing. This was partly an attempt to "people" history in the absence of structures, but also an attempt to enliven what was essentially a park space with things to see and do. It too had to be an attraction. As the Forks management plan noted in 2007, "the array of entertainment and leisure opportunities provided by adjacent landowners makes the communication of historical messages to visitors more challenging – there is much competition for their interest and attention. The Parks Canada Property is only part of this much larger Winnipeg attraction."[28]

This approach was not without its critics. The chair of the Historic Sites and Monuments Board felt this approach reduced a national historic site to "a staging area for 'heritage' events" (the use of scare quotes around "heritage" says it all). The Manitoba Heritage Council and Historic Resources Branch read it as a centralist slight that signalled the Forks' lesser status in the national historic sites system dominated by major projects in eastern Canada. (The province had just funded a reconstruction of Fort Gibraltar on the other side of the Red River, so it would be underwhelmed by a minimalist "sense of place" approach.) And the public in Winnipeg seemed to draw the (entirely valid) conclusion "that the activities are unfocussed and not linked."[29] Against the wider concern over commercialization, it was not clear how the historic site was supposed to be teaching Canadians about the Forks.

What it does do is sustain the Forks as a whole, as a core public space for Winnipeg. The Forks supplies a smaller version of the extensive riverine parks curving through other Prairie cities, like Calgary, Saskatoon, and Edmonton. The Parks Canada section provides a green respite from the bustle of the market, shops, and cafés; the Forks North Portage Partnership describes it "a place which is safe, clean, green, affordable, diverse, connected, and attractive." (Tellingly, there is no mention of history.)[30] The ungenerous reading is that it gives a lively commercial complex an appealing veneer of public education. At the very least, it complements but does not compete with the commercial attractions. So the National Historic Site serves as "a peaceful stretch of land dedicated to quiet contemplation of times past," and as "an oasis of calm and an opportunity to commune with nature in the centre of a busy urban environment."[31] The Variety Heritage Adventure Park, an elaborate historically themed playground structure for children, is the sole "attraction" on the Parks Canada property, and it is popular because it serves what the public wants from this *section* of the Forks complex: family-friendly, safe, overtly recreational, and just quietly educational.[32]

As a hybrid of park/attraction/designation, the Forks embodies several patterns at Canada's historic sites. Like other protected places, it exists on an axis of preservation versus use. The more popular it is, the more worn it is. Historic sites, located closer to many Canadians than national parks, have often been packaged as useful green space for scenery and recreation. The town of Annapolis Royal used Fort Anne as a baseball diamond at the turn of the twentieth century; the Windsor Golf Club won a fifty-year lease on Fort Edward in 1924. As we have seen at Grand Pré, the romance of the setting – "in these green hills, aslant to the sea, no change!" – was a major driving force in tourism to the land of Evangeline. Picnic tables are still scattered around the earthworks at Fort Anne.[33] While the tradition of marketing national parks and historic sites is as old as the system itself, in recent decades it has become rather more … interesting. The re-designation of Parks Canada in 1998 as an agency (rather than a branch of a federal department) resulted in a quasi-corporate identity (with, for example, a CEO rather than a director) and a practice of marketing its properties as multipurpose spaces that can be rented for

functions such as weddings and conferences. Each property offers a
distinctive venue without much constraint in terms of use. The Forks,
for example, offers brides-to-be "an historic place for your historic
day," with the backdrop of the Red River and St Boniface Cathedral
for their ceremony. It is not the historic or ecological meaning that is
as valuable as the setting. It is, again, useful space.

Environmental History at the Forks?

But the Forks would be the ideal spot to foreground the environmental history of western Canada. Much of its interpretation is terrestrial
in nature, discussing the archaeological layers of soil, or movement ra-
diating onto the prairie by rail and road. But there could be much
more said about the politics and problems of urban rivers.[34] I was
stunned to find older plans for the Forks that suggested precisely this,
more than twenty-five years ago. The Red and Assiniboine were recognized as an important means of transportation for canoes, York
boats, and steamboats. Most early plans hoped to see renewed boat
traffic. But they went further. Between 1989 and 1995, in the first
years that the site was open to the public, a series of proposals consistently and explicitly imagined a public message of environmental
history and environmental engagement that has largely vanished from
the Canadian political arena.

   In 1989, *Time and the River: A Conceptual Interpretive Plan for
the Core Area Riverbanks of Winnipeg* opened with the salvo, "Do
contemporary Winnipeggers have many ways in which to gain an
awareness of those elements affecting the river or of how the river was
used in the past?" Its periodization was entirely river-focused: from
glacial lake bottoms, to river lot farms and industrial docks, to city
parks and sanitation. It emphasized human dependency on riverine
environments as well as human alteration *of* these environments,
as with the contrast between native diversity in a river bottom and
European monoculture in a Victorian park, or the "stabilizing" of the
river and flood management. *Time and the River* was the only study
to evaluate historic resources according to "a riverbank focus" *and*
locate themes of riverine and water-based environmental history

throughout the city: proposing a discussion of urban sanitation on the former site of a city waterworks plant, for example, or a buried creek under the city's exhibition grounds. It is an excellent example of public environmental history, and one unlike almost any I have seen.[35]

Four years later, the *The Forks Heritage Interpretive Plan*, commissioned by the Forks Renewal Corporation, also conceived of history as "an interactive relationship between people and the natural setting." Although more landward in orientation, it argued that these Euro-Canadian settlement patterns and Winnipeg's lot divisions were determined first and foremost by the rivers.[36] But most surprising of all is a report authored by Parks Canada itself in 1995. *The Forks National Historic Site: Historic Themes* divided the history of the Forks into eight fairly conventional chronological periods – "the competitive fur trade period," "the Forks and immigration" – but then added a ninth theme of "Environmental Citizenship." Citing Canada's 1990 Green Plan and its "call for immediate environmental action" and "an environmentally literate society," *The Forks National Historic Site: Historic Themes* explicitly called for programming at the Forks that would "range from the local urban setting to the global ecosystem, from the environments of the past to our current situations. This information is to be associated with the experiences of the site's visitors and include suggestions for positive action."

Ten Principles of Environmental Citizenship
1. Care for our planet's air, water and land.
2. Reduce, recycle and reuse.
3. Control our use of non-renewable resources.
4. Use our renewable resources wisely.
5. Safeguard our planet's diversity of life.
6. Foster appreciation for cultural and historical heritage everywhere.
7. Be sensitive to the environment in our financial decisions.
8. Foster education and our understanding about our planet and our role in it.
9. Work for a healthier environment personally and in our communities.

10. Urge our government to work with others for a healthy
    global environment.[37]

It is, quite simply, impossible to imagine Parks Canada or any federal
agency issuing such a statement today. What was it about the early
1990s that made this possible? Was it having a federal policy (the
Green Plan) that directed the government to prioritize environmental
values? Was it because the site still anticipated some public funds?
Was it the more (L)iberal political climate? And could we revive this
challenge of civic engagement?

What if we consider the relationship between the city and the rivers
anew, especially the ways in which we are implicated in their present
shape? For example, could tours or installations at the Forks discuss
the location of a city on an ancient lakebed? Manitoba's lakes, rivers,
and rich soils are all remnants of glacial Lake Agassiz, which drained
about nine thousand years ago. This historicizes surface water: it reminds us that shorelines and river channels have changed over time.
Geological studies suggest the river channels once flowed through the
Forks. More importantly, people change rivers to *allow* for cities:
draining wetlands and sloughs, burying creeks, filling in riverbanks
for building.[38] In other words, what we see at the Forks and the downtown is not next to the river, but sometimes, on and in it.

Or what of the need to supply urban Canadians with nature even
on a frankly non-natural river? Six kilometres upriver, Winnipeg had
created one of the earliest urban parks in Canada, when Frederick Todd
designed Assiniboine Park in the style of Frederick Law Olmsted, with
lawns, winding avenues, and extensive formal gardens.[39] The Red
River Floodway, completed in 1968, only a few years before serious
consideration of the Forks redevelopment, was one of the largest
megaprojects in Canadian history. It designed to divert the river *away*
from the city. As an "opportunity to commune with nature" the National Historic Site is a somewhat disingenuous exception along a
heavily engineered corridor. With so many North American cities located in floodplains, we *assume* their control – a control that permits
our cities to exist. As one historian has noted, we "tend to think of rivers
as canals rather than meandering and dynamic streams."[40] Would

remembering their flooding history cultivate some humility? How does our desire to draw people *to* the river here coexist with our desire to hold *back* the river? What does it say about our attitudes toward nature when we refer to a dyke as a "Primary Line of Defence"?

And what does the story of the Forks tell us about how we value riverine nature? In the latter part of the nineteenth century, the Forks was considered "wasteland" – flooded, swampy, unstable – but never useless. As Jennifer Bonnell has shown of the Don River in Toronto, a difficult river could serve as an urban fringe, a place to locate industrial plants and other undesirable elements (immigrants, prostitutes) still necessarily for capitalist modernity. If Winnipeggers in the 1980s found the CN East Yards unsightly, to their predecessors in the 1880s the greater blight would have been land left empty, unused, and unprofitable.[41] Can we reclaim other "waste" lands?

The story of the Forks is, in fact, a hopeful one. The Forks made a significant portion of an urban waterfront accessible and enjoyable and, it must be said, useful again. It also expanded our concept of historic site. Only in recent years have we begun to assign heritage designation to the large-scale industrial complexes so essential to the development of Canada. Here is one such site both recognized and reclaimed, a lesson of some value to cities across North America. By making use of existing railway structures (and removing the unusable ones), it demonstrated how a historic site could encompass native ecologies and twentieth-century industrial plants in a post-industrial framework. It grappled with questions of archaeology, habitat restoration, and other elements of historic landscape that characterized the evolving nature of historic sites in the 1980s and 1990s. And it showed how historical space can exist within urban realities, rather than as a time capsule set apart. There is much to recommend the parkland model of historic site, but we need to ensure it does not become merely a passive or convenient complement to the urban landscape. A historical park can remind us of the natural world and waters running underneath, through, our cities, and our history, from sea to sea.

CHAPTER 5

Nature's Gentlemen and a Nation's Frontier: The Bar U Ranch

> Probably every country and every area has a romantic period in its
> history; in southern Alberta the early ranching years must qualify
> as its golden days.
> – Sheilagh S. Jameson, "Era of the Big Ranches" (1970)

Standing on the ridge above at the Bar U Ranch National Historic Site,
it's easy to believe in the romance of this place and its history. The
Rocky Mountains gleam white in the distance. The wind rushes in
your face, bringing the scent of prairie grass. The cottonwoods cluster
along Pekisko Creek. Grasslands – green in summer, tawny in fall –
arc smoothly up from the creek bottom. Red wooden stables line a dirt
road. There's no sight or sound of modern occupation (if the highway
behind you is quiet). It's all quite unbearably beautiful. The historic
site is a wonderful glimpse of a prairie past, with the most extensive
collection of *in situ* resources of all the sites discussed in this book. It
also holds the most ambitious, and controversial, environmental story,
and the one that reveals the most about ourselves as a country.

The Bar U was established in 1882 when the federal government
granted extraordinarily large leases of land to a handful of investors
from eastern Canada and Great Britain. The climate and grasslands
of the foothills were well suited to herd grazing, but like other elements of national policy, these leases were designed to make the
northwest profitable and assert Canadian sovereignty against American interests. Though most of the larger ranches were broken up for

5.1 The Bar U Ranch, 2016. (C. Campbell)

farms, the Bar U remained a working ranch until the 1950s. It was thus an ideal choice for designation in the 1980s when Parks Canada sought to diversify its western sites beyond the fur trade and Mounted Police posts.

Despite its relative youth, the Bar U preserves an older view of the frontier: simultaneously a wilderness (expansive, with inexhaustible potential) and a garden (productive, habitable, well-governed). This is appealing, but not entirely accurate. The ranch was part of an emerging transcontinental network of industrial agriculture, a capital-intensive project of production, shipping, and processing; and even with a "naturally" rich grassland, it required enormous infrastructure. Nor was it alone on that frontier. Ranchlands overlap with emerging landscapes of oil and gas exploration, and tendrils of urban growth. We need, then, to ask why the Bar U Ranch National Historic Site offers us a frontier separate from industrial modernity when, in fact, it is not. We need to ask what it permits us to believe about our tradition of a resource economy and our commitment to that tradition.

"Strictly a Business Operation": Ranching in Western History

The history of ranching has been told as a story of freedom enclosed, a pastoral culture taken over by corralled production, hard-working cowboys on the open range displaced by stockyards and fenced lots. But ranching had always been part of a project of colonization, capitalism, and modernity. It was, in fact, corporate capitalism that created the environment for cattle in the first place. Buffalo on the northern plains, when processed into pemmican, became an indispensible food source for not only the rival fur trade companies but also other agents of colonial power: explorers, scientists, missionaries, and settlers. We can also see in this capitalized buffalo economy a forerunner to industrial ranching: an animal processed into commodity, exported for distant audiences and external profits.[1] The buffalo hunt cleared the grasslands of its native grazers, and decimated the plains peoples and Métis who depended so completely upon them. Alfred Crosby called this ecological imperialism: colonizing new territories by removing indigenous species and cultivating new ones in their place.[2] There is no better and sadly ironic illustration of this than the fact that one of the first markets for ranched beef in western Canada were First Nations who had lost their food source with the buffalo and depended increasingly on supplies promised by treaty in case of general famine and "the calamity that shall have befallen them." The ranches in southern Alberta occupied this geography of displacement, surrounded by reserves carved out for the Tsuu T'ina (Sarcee), Blackfoot, Crow, and Blood, who – another irony – supplied essential skilled labour as ranch hands.[3]

As it displaced indigenous ecologies, livestock signalled permanence and self-sufficiency for nascent settler colonies in the northwest. The Hudson's Bay Company imported stock from the Missouri Territory in the 1820s for its headquarters at Fort Garry (Winnipeg), and later exported cattle southward, despite the difficulties of the long Midwest drives.[4] Cattle were also a key food source for mining camps in both countries, as in the gold rush in British Columbia in the 1850s. But as one settler near Fort Garry complained to geologist H.Y. Hind in 1859, "Look at that prairie; 10,000 head of cattle might feed and fatten there for nothing ... but what would be the use? There are no markets."[5] The wealth of the interior could only be harvested with the infrastructure of the industrial age. As David Breen put it bluntly in his early and influential work *The Canadian Prairie West and the Ranching Frontier 1874–1924* (1974),

> Cattle ranching as it developed in North America in the early 1880s was strictly a business operation and in this sense is a product of the Industrial Revolution. The industry depends on large urban markets as well as a massive processing and transportation infrastructure to link producer and consumer … In the broadest sense the phenomenal expansion of the ranching frontier in the late 1870s and early 1880s is as much an expansion into the working-class areas of London, Manchester, and cities of the American seaboard as into the grasslands of the northwestern plains.[6]

From the outset, easterners entwined desire for that wealth, and faith in that infrastructure, with their sense of Canada as an emerging nation-state. The right to profit from the interior represented future prosperity and promised sovereignty. Cattle ranches represented a harbinger of permanent settlement, one of a series of strategies – along with building a transcontinental railway, gridding the prairie for farming, and establishing a mounted police force as the instantly recognizable symbol of federal authority – to acquire, domesticate, and occupy the west. In the 1860s and 1870s, Ottawa was highly anxious about American designs on its "empty" western territory. A 1869 cartoon by J.W. Bengough depicts a buff, uniformed young Canada booting an unsavory American off the steps of Dominion House, while John Bull looks on approvingly, saying, "That's right, my son. No matter what comes, an empty house is better than such a tenant as that!"[7] Large ranches would use eastern capital to hold western space, to occupy Canada's "empty house" until the prairie could be peopled with farming families and wheat.

The Wealth of the Foothills: Large Lease Ranches

In 1881, Matthew Cochrane, a senator and industrialist from Montreal, proposed a large ranching project in the foothills of the Rocky Mountains, and asked John A. Macdonald's Conservative government to "assure" to him sufficient lands "in view the large amount of Capital intended to be embarked therein." Accordingly, a ministerial order amended the 1872 Dominion Lands Act to allow for leases of up to one hundred thousand acres, for a period of twenty years with an annual fee of ten dollars per thousand acres, or a penny an acre. This was agriculture on an entirely different scale from the homestead or family farm, and looked much more like the practices of leasing Crown lands for mineral or timber harvest and export.[8]

Those who established these "large-lease" ranches likely recognized the industrial framework in part because they themselves were already heroized as captains of industry and builders of empire. As one commentator wrote in 1905, "You cannot start ranching with nothing but faith and muscle, as you can wheat-growing. A nag and an old cow is nothing of a beginning. Ranching means money."[9] Cochrane was a prosperous footwear manufacturer who bought purebred livestock; the North West Cattle Company, which adopted the Bar U brand, was headed by Frederick Stimson, a prosperous livestock dealer from Quebec, and backed by the wealthy Allan family, also of Montreal, whose shipping and banking interests marked them as among the richest men in Canada. They brought the necessary investment capital, but also political influence and knowledge of transatlantic markets. And so the North West Cattle Company began with two leases for 147,000 acres – it would eventually expand to include nearly seven townships of deeded and leased lands, or nearly 158,000 acres.[10]

The location of the large ranches was crucial. They would be deep in the interior, near the Montana border (to boot the Americans off the front step), but close to several Mounted Police posts and the projected route of the transcontinental railway, which reached Calgary in 1883. The foothills promised shelter, water, and a sympathetic climate, with the mild winds of the chinooks punctuating the prairie winter.

Most importantly, the foothills promised grasses: notably rough fescue, a resilient, deep-rooted grass that cures on the stalk in fall, making it an excellent winter food source for cows and deer. But ecological riches gave rise to ecological myth: that this "prairie wool" would nourish cattle foraging freely over the long winter.

We can see characteristic attitudes toward the frontier at play in the large-lease era, as an empowering trove of resources, with little concern for cost. It assumed free land and high prices for beef, both of which were lost by the 1880s. (Some ranches like the Bar U survived by purchasing much of their leased land, which gave them the necessary space but further bound them to networks of banking and capital.) It assumed cattle – any number of cattle – could be pastured freely with little care or attention, which was impractical in the face of predators, disease, and crippling winters that regularly killed tens of thousands of head. It assumed prairie grasses could sustain cattle over weeks' of traumatic transport by rail and ship to British cities. In short, it assumed an inexhaustible commons, a natural resource that could be harvested with a reasonable profit margin. It recognized the worth of the foothills in the prevailing economic paradigm of resource expansion, but it did not readily accept the limitations of either place or paradigm. Warren Elofson, one of the leading scholars of ranching history, notes the disconnect between metropolitan thinking and environment at play here: "Pasture management generally should be seen as a frontier circumstance in the sense that, to some degree at least, it related to the ranchers' ignorance of conditions in their new land; and it occurred throughout the entire North American West."[11]

Thus, the large ranches, for all their advantages, could hardly be considered savvy business ventures. Instead, by the early twentieth-century a more sustainable model of mixed farm-ranch appeared, keeping smaller herds fenced more closely, providing winter shelters, and, critically, growing hay and other feed. This seemed to recognize the limits of natural bounty on the one hand and the responsibility of human cultivation on the other – to replace a remote corporate directive with a more sympathetic, localized practice.[12] However, these adaptations only fed, literally, into the growing infrastructure of transcontinental, industrial agriculture. The farm-ranch at one end of the rail line now required a feedlot at the other. The Bar U was

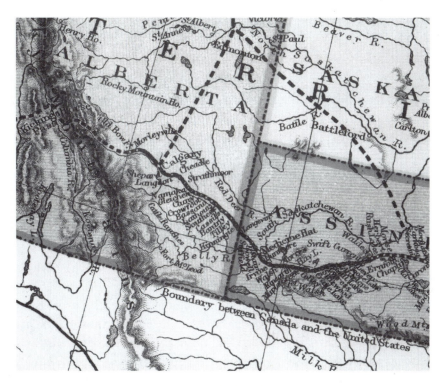

5.2 Department of Railways and Canals, *The Dominion of Canada: A Reduction of the Map Prepared & Issued under the Direction of the Minister of Railways & Canals*, 1882. Published in John Douglas Argyll, *Canadian Pictures Drawn with Pen and Pencil with Numerous Illustrations from Objects and Photographs in the Possession of and Sketches by the Marquis of Lorne* (London: Religious Tract Society, 1883). (Courtesy of the University of Manitoba Archives and Special Collections)

unusual in that it remained a large ranch, but not unusual in its integration into other branches of the agricultural industry. In 1902, former ranch foreman George Lane bought the North West Cattle Company ranch, renamed it the Bar U, and began breeding Percheron horses for transatlantic buyers; twenty-five years later, it was sold to meat-packing magnate Patrick Burns, whose business interests extended through the cattle, sheep, and dairy markets. Furthermore, beef production typified the weaknesses of a resource economy: dependent on export markets, whether British or American, and accordingly, vulnerable to protectionist tariffs and global price fluctuations.[13] It is, in that sense, a cautionary tale for the Canadian economy as a whole.

Remembering the Range

But we have not seen the ranching frontier as such a tale. Instead, for over a century it has been cast as "a romantic period" of the west's "golden days." (Jameson, herself daughter of a rancher, as well as author and chief archivist of the Glenbow Foundation in Calgary, was typical in her characterization of western history.) As American historian David Wrobel notes, "The mythic West is so much easier to deconstruct within scholarly contexts than it is to exorcise from the public consciousness because, in part, of the efforts of older generations in playing up the adventures and glories of the frontier past. The powerful legacy of pioneer reminiscence helps explain why the West's mythic veil is so difficult to remove."[14] The industrial features of nineteenth-century ranching were almost immediately obscured by a popular nostalgia for the mythical free range, equally alluring to east and west, to tourists and residents. Indeed, this version of history, and landscape, was scripted even before the end of the nineteenth century, and reinvented with each generation. Theodore Roosevelt proclaimed in 1888 that "The great free ranches, with their barbarous, picturesque, and curiously fascinating surroundings, mark a primitive stage of existence as surely as do the great tracts of primeval forests, and like the latter must pass away before the onward march of our people."[15] Twenty-five years later, L.V. Kelly mourned that "the range is gone forever, cut up by the fences of the farmer, and the railroad ... The ranches have gone, the open rage is enclosed in fences, the wild cattle no longer roam at will across the broad sweeps of the prairies."[16] Then in 1927, famed cowboy artist Charlie Russell summarized the standard memory of events:

> "The cow ranches that I knowed," says Rawhide Rawlins, "is nothing like them they're running to-day. In the old days, it wasn't much, only a place to winter. They were on a stream or river bottom. The buildings were made of what the country gave – logs, either cottonwood or pine in the North. They had one house – maybe two, with a shed between, a stable, and a pole corral. All of these buildings were dirt roof, some had no floor but ground. There were no fences, not even a pasture ... Most of

the cow ranches I've seen lately was like a big farm. A bungalow
with all modern improvements, a big red barn that holds white-
faced bulls an' hornless milk cows" …

"The cow-ranch to-day," says Rawhide, "is a place to make
money to go somewhere else."[17]

What Russell chose not to mention, or remember, was that the cow
ranch had *always* been a place to make money that was going somewhere else.

The centerpiece of this nostalgia in Canada became the Calgary
Stampede, founded and funded in 1912 by men known as the "big four"
of the Alberta cattle industry, including two owners of the Bar U. As
a competitive display of cowboying skill, the Stampede capitalized on
the steady nostalgia for a lost open range and offered a new, revenue-
generating platform for the heady boosterism of a western city. It also
became a cornerstone for Albertan identity, in a way that allows residents of the financial capital of western Canada, thanks to its role in
the oil and gas industry, to connect with an imagined rural past:

For ten days in July, men and women from all walks of life don
tight-legged jeans, wide-brimmed hats, and ornate leather boots
as they cavort through they city's streets, bars, and midway. The
"world's greatest show outdoors," as the Calgary Stampede
modestly describes itself, is in fact a celebration of some ill-
defined "frontier democracy" that is located vaguely in Calgary's
distant past in a world where social equality and economic opportunity ride spiritedly across the prairies. It is a powerful myth,
promoted and perpetuated by virtually all of Calgary's public institutions, including the city's media, museums, libraries, schools,
and civic administration. Like all myths, it contains an element
of truth. Nevertheless, this Turnerian vision of Calgary's history
remains an invention of the past.[18]

In fact, Alberta's vision of the past looks much more like Calgary
than it does a cowboy ranch. Most of the province's designated historic
places date between 1905 and 1913, after the achievement of provincial
status, and an influx of settlers and capital prompted a building boom.

Handsome stone and brick banks, warehouses, and four-square houses are all durable and aesthetically pleasing markers of wealth and prosperity. In other words, historic sites have preserved best the arrival of industrial capitalism on the Prairies. At the same time, local and community histories and events tended to be more rural in focus, celebrating the "vanishing pioneer past."[19]

Ironically, it was another era of growth that reoriented Alberta's approach to historic sites. The 1970s saw a rapid expansion of provincial heritage projects, coinciding precisely with the greatest development boom in the province's history. The explosion of the oil and gas industry, and the spinoffs in highway and housing construction, created a "development juggernaut" that both unearthed and endangered historic resources, particularly in rural areas. The Cochrane Ranch, to this point the leading contender for designation as a historic site to represent Canada's ranching history, was caught between suburban projects and highway expansions – ironically, the expansion of Highway 22 west of Calgary, which was later billed as the Cowboy Trail. The Parks Canada historian tasked with studying ranch sites conveyed the new landscape when he noted of the Cochrane site, "This road system, which is now complete, along with railway, power lines, Town of Cochrane development, lumber mill, gas plant and other modern intrusions present an environment in which it would be very difficult if not impossible to interpret the era of big ranches."[20]

It also, unfortunately, hardened jurisdictional lines. In 1972, Alberta's Environment Conservation Authority admitted that "inflexible attitudes" in Edmonton and Ottawa over, in particular, mineral rights had left Alberta as the only province in Canada without a National Historic Park. At the same time, the province recognized the economic value of tourism *as* an industry, and was ready to develop historic sites to generate revenue in underdeveloped rural areas. New sites in the 1970s and 1980s were located along major highways but away from the economically healthy urban centres of Calgary and Edmonton, and the perennially popular national parks.[21]

Ranching *in situ*: The Bar U National Historic Site

In 1968, the Historic Sites and Monuments Board of Canada recognized ranching as an important part of Canada's history, and recommended acquiring a ranch (specifically, the Cochrane) in order to interpret this theme to the public. Parks Canada then commissioned a survey of ranches in southern Alberta, which examined seventeen potential candidates. The American National Park Service, which had done a similar survey fifteen years earlier, was in the process of developing a ranch in western Montana. (The Grant-Kohrs Ranch was also located in the foothills, five hundred kilometres due south from the Bar U.)[22] The Parks Canada survey praised the Bar U's extant collection of buildings as evidence of its business success: "The Bar U was splendidly managed and handled from the start," Harry Tatro wrote. "One cannot view it without being impressed with the exceptionally well-kept appearance it presents."[23] Eventually, the HSMBC revised its recommendation to prefer the Bar U, recognizing the unusual amount of *in situ* resources and, more usefully for interpretation, well-kept buildings (although not the ranch house, which had burned in 1927). In 1989, Parks Canada acquired a property of about 148 hectares (367 acres) – sizable for a historic site, but still only one-quarter of a percent of the ranch's size at its height! The complex site included an unusual thirty-eight buildings, including ranch office and post office, a harness repair shop, blacksmith shop, cookhouse, horse barns, and coal shed.[24] After a few years of management planning, it opened to the public in 1995.

The Bar U National Historic Site celebrates a story of corporate success and longevity, and a Canadian presence in a competitive international market. After all, it was available for designation largely because it had weathered (literally) meteorological and market shocks more successfully than the other large ranches. So Parks Canada stresses the role of ranching in the Canadian economy; the Bar U as a model of corporate ingenuity in Canadian history; the men (Frederick Stimson, George Lane, Patrick Burns) who acted as spokesmen for the industry; and the ranch's longevity over time. The Friends of the Bar U are even more explicit in their admiration of its business success: "Bar U cattle literally fed the world ... While other large Alberta ranches succeeded for a time only to go out of business, especially after the killer winters of 1886 and 1906, the Mighty Bar U [*sic*] persevered to eventually become a kingpin in a business empire that included a variety of ranches and farming enterprises, as well as meat packing plants and flour mills."[25] Of course, no one expects a national historic site, or public heritage more generally, to commemorate major business *failures*. But such a triumphant story sits uncomfortably against what we now know from environmental history, in the long-term ecological costs of industrial ranching, and the pattern of unsustainable resource economies in the western interior.[26]

This is particularly odd given the attention that was given to ecology and landscape in the site's planning. The Bar U was one of the first projects to be undertaken under the 1990 Green Plan, a federal White Paper that came with a $3 billion budget at a time when Parks Canada was situated within the Department of the Environment. But the Green Plan itself carried the ambivalence toward "frontier" that pervades the Canadian psyche. It certainly did not critique the premise of a heroic resource frontier; indeed, it promised to "fill the major thematic gaps within the national historic sites system," in which *"priority sites will represent most of the key resource development industries and settlements"* in Canada's history. Scholars have noted that the *lack* of opposition to the plan by the business community (especially, for our purposes, in Alberta) and the provinces demonstrates its essentially (neo)conservative philosophy, one that avoided regulatory regimes in favour of "sustainable development."[27]

Nevertheless, the Green Plan reminds us that the Bar U was developed in an era of public environmental concern, as well as environmental impact assessments, commemorative integrity statements, and cultural landscape study. Management documents written in the 1990s and 2000s stress the relationship between range ecology and ranching history. They devote a great deal of time to cultural landscape elements (fences, corrals, dispersed ranch buildings, pasture and cultivated fields) and natural features (native grassland, sheltered creek valley); the way in which ranching practice depended on these features; and how ranch buildings were aligned with the topography. There is also some concern over the intrusions of agricultural settlement (fences, roads, farmsteads), highway construction, and a pipeline corridor; the integrity of remaining patches of native fescue grassland; and the site's (especially the creek's) role as habitat for a various bird and fish species. In short, there is a sense of land, water, and weather that marks this as one of the "newer" generation of historic sites.[28]

Still, these ecological concerns were site-specific, not systemic. They identified the particular features of the Bar U, but did not consider them representative of wider attitudes toward natural resources. Such ideological inertia explains why Parks Canada's position on environmental engagement seems to have actually eroded. The original management plan for the Bar U in 1995 promised to "demonstrate the principles of ecological sustainability in a rural setting [and] promote environmental citizenship and practice environmental stewardship." Remember that in 1995 Parks Canada also added "environmental citizenship" as a key theme to the Forks at Winnipeg. *Ten years later, no such statement appeared in the revised management plan for the Bar U.* Instead, "Staff, partners and visitors will be told of the importance and the challenges of environmental stewardship at the Bar U."[29] The revised language scripts a passive role for visitors, rather than seeing them as agents of environmental engagement. The euphemistic term "challenges" softens any sense of urgency, and yet also suggests the problems are too great for resolution. I am reminded of Matt Dyce's description of proposed interpretation on the Athabasca Landing Trail in boreal Alberta, in the 1970s: "This new approach to the natural environment positioned nature somewhere in between an object of preservation, a renewable resource to be harnessed for the economic development of the region, and a territory of control."[30] We can care for places of significance, but not want much to change.

Ranching as Romance or Industry?

I want to stress this is not exclusively, or even primarily, the fault of Parks Canada. It is the result of us choosing what we want to see. The most powerful experience of the environment at the Bar U is a scenic and sensual one: the unobstructed view to the Rockies, the breeze in the cottonwoods, the sound of horses in harness. The view is crucial, for its beauty and for conveying notions of boundless space so important to the period of national expansion. This is, after all, meant to be the frontier. The particular value of this site becomes clear when compared to that of the Cochrane Ranch, now surrounded by the kind of housing developments and major highway routes that characterize much of suburban Calgary. Despite the messages of corporate organization and longevity, *and* the fact that it is categorized under "extraction and production" in the national historic sites system, the Bar U does not present a landscape of industrial modernity. The ranch, quite simply, looks nothing like the mines, railways, canneries, or brickworks that we associate with "industry," so there is almost no way for a visitor to connect this stunning locale with an economy of extraction and harvest.[31] (And because of the separation of ranch, abattoir, and store, we rarely connect a plastic-wrapped steak to its origins.) Indeed, it allows us to *escape* the environmental costs of such an economy, and therein lies the crux of the problem. While plenty of historic sites celebrate capitalist values of ingenuity, growth, and prosperity, the Bar U seeks to offer a glimpse of a frontier left intact and unharmed by the actions these values inspired.

There is a long history of characterizing of ranching as somehow different from other resource frontiers. Ranchers will point out that grazing keeps the fescue healthy. Cattle, like bison, suit the grasslands. Cowboys are perceived not as workers or employees but as uniquely autonomous, and with the charge of raising *living* beings rather than harvesting an inert resource. They were (and are) "Nature's gentlemen," with an inherent dignity and moral code.[32] The cowboy as a "knight of the frontier" is usually credited to American writer Owen Wister, writing in *Harper's Monthly* in 1895. But Canadians have had a particular investment in the trope, as a way of affirming their British heritage on a (North) American frontier. Ralph Connor's 1899 novel *The Sky Pilot*, set in the Alberta foothills, features "the Noble Seven," a group of well-born gentlemen-ranchers who embody camaraderie and chivalry, and consequently form "the only social force Swan Creek knew."[33]

But even as Canadian mythology invoked the rancher's nobility to suggest a distinct Canadian frontier, the trope nourished a common political ideology across the border. If cowboys have an innate sense of virtue, and a unique knowledge of, even rapport with, the land, then they should be entrusted with its government. This is the argument that has been made again and again as western communities challenged eastern authority for control of land and resources, whether the "Sagebrush Rebels" clamoring for more access to federal lands in the 1970s or opposition to the proposed national park in the British Columbia interior.[34] Scholarly and popular histories alike present ranchers as denizens of wilderness with an empathetic understanding of nature and cattle, who reluctantly managed rather than tamed the land, dismayed at the earth ploughed by homesteaders. "It is a challenge to make the best use of the good earth without abusing it," one rancher told author Andy Russell, himself a prominent conservationist.[35] Plans for southern Alberta consistently refer to the historic stewardship of its ranchers, who bring a sense of tradition and responsibility, especially in contrast with "larger urbanized society … growing more and more estranged from the realities of livestock raising."[36] Some have adopted practices such as rotational grazing, and have led public protests against exploratory drilling in the foothills, for its disturbance to fescue. And they have put their own land on the table, through private land donations and conservation easements. This, I would argue, *does* model the environmental citizenship called for in the site's first management plan.[37]

But as Donald Worster has pointed out, both Canada and the United States framed their wests in an ideology of development, as exploiting the land for wealth in some way. Both have constructed narratives that legitimize this ideology: "Canadians have liked to describe themselves as gathering staples from vast northern country while becoming spiritually part of what they exploit … A nation bringing law and order to the continent while obeying the laws of nature."[38] True, ranching did not seek to "prove up" (improve) or cultivate nature in the same way as farming; it did not insist that "rain follows the plow." But it did require enormous investments in irrigation, fencing, haying, and stock, and a sympathetic legislative framework. Ranching histories note wear and tear as early as the 1880s, when the much-touted native grasses were stripped by too-large herds, then turned over with the sod and crowded out by wheat and hay. Watering holes were a source of conflict. And cattle died by the thousands. It does not damage the land as other industries have done, but it is nonetheless part of a capitalist system that displaces, and profits from, natural wealth. *Rancher* in French, after all, is *exploitant de ranch*.

It is thus both fitting and problematic to see cowboy imagery used to ennoble other resource industries – in particular, to convey western sovereignty in the politics of oil. Alberta has a long history with fossil fuels, a history that overlaps almost exactly with that of ranching in the area. By the 1880s and 1890s, coal was mined north and west of the Bar U, into the Rocky Mountains (Pat Burns owned a mine at the headwaters of the Sheep River), even in Banff National Park. Gas fields in the southeast of the province were so extensive that Medicine Hat was dubbed Gas City.[39] But it was black gold that would anchor Albertan identity in the twentieth century. Drilling enthusiasm spread across the foothills in 1913 when oil was discovered at Turner Valley, which then became the largest oilfield in Canada. When Imperial Oil's well at Leduc, south of Edmonton, blew in 1947, it transformed the landscape of oil production in North America. And today, the Athabasca oil sands (or tar sands) have become the dominant icon of the degradation caused by fossil fuels – both the processes of extraction, and the climate change produced by emissions. In each case, the energy industry has required the same kinds of things (large amounts of investment capital, heavy machinery, an imported labour force) at significant environmental cost.

Whether coal, hydro-electricity, or oil, energy revenues have long been the centerpiece of provincial revenue, and accordingly, of provincial power within Confederation. And to symbolize that power, Alberta consistently adopted the cowboy image to convey its desire for greater independence from Ottawa. Large-lease ranches represented an arm of a nineteenth-century national policy; but ironically, the cowboy became a symbol of Alberta's rejection of a twentieth-century one (the National Energy Policy of 1980). In other words, the province, and the oil industry, borrows the most benign, and popular, image of western resource harvest as its avatar. Their stories are intertwined, even in the physical proximity of historical sites. The province declared Turner Valley a provincial historic site in 1989; the Historic Sites and Monuments Board named Turner Valley a national historic site in 1995. But it has required years of remediation for the hydro-

5.3 Department of Mines, *General View of Turner Valley Showing Imperial Oil Separator in Foreground*, 1928. (Courtesy of Library and Archives Canada)

carbons and heavy metals, and is not yet open to the public.[40] And it is only thirty kilometres north of the Bar U.

Frontiers of Opportunity?

Here is the crux of the matter. The Bar U is a protected place, but it is not an island unaffected by larger environmental practices. It is a stone's throw from a massive oil field. It is a part of western Canada that has suffered significant drought due to climate change. It is an hour's drive from Calgary, a city with some of the worst urban sprawl in North America. The foothills are increasingly desirable for acreages and residential subdivisions, which threaten the patches of fescue (the provincial grass of Alberta since 2003!) by introducing non-native grass species. And yet, it is that very environment that makes it so desirable: "Today's creative western developers … promise affluent metropolitan refugees the prospect of an escape from the postmodern angst accompanying overcivilization. Nonetheless, today's western real estate advertising does echo the promotional past – with its emphasis on frontiers of opportunity – in presenting these places as the last available frontiers of landed paradise."[41] Sheltered, literally, in the Pekisko Creek valley, we can choose to see the Bar U as a sliver of historic paradise, not as part of an inhabited landscape. The ranch lets us picture the frontier – and by implication, the continental interior – as whole and hale. It provides a balm to our other image of Alberta and frontier: the aerial views of the Athabasca oil sands. It lets us have our environmental cake (enjoy it aesthetically) and eat it (profit from a resource economy) too.

Policy encourages this fiction. We compartmentalize land use, whether through zoning or designations, and isolate the impacts, so we rarely need to acknowledge the cumulative effect of our history in the land. As Douglas LePan once observed of the archipelagic Georgian Bay, "angels alone would see it as whole and one."[42] We can hardly congratulate ourselves on protecting our "national treasures" in national parks like Banff or national historic sites like the Bar U when they are surrounded by oil fields. The tar sands, Calgary's sprawl, and the Bar U co-exist in such close proximity *because* they are products of the same national trajectory, the same extractive logic of modernity, the same foundations that underpin Canada's economy today. We need to see this landscape, this history, *in toto*, to recognize its cumulative effect, and just how fully it is woven into the fabric of our national imagination. My concern is that the Bar U and places like it act as an alibi for our unsustainable, extractive national economy. History should be a cautionary tale to guide better choices, not absolution for our behaviour. We need to see the ranch as part of the region, but more importantly, as part of an industrial mentality that continues to define how we approach our resource frontiers. The "free range" era was never truly free. No resource wealth is.

humans, historic sites commemorate "second nature," or the land-scape produced by human intervention in natural systems.[2] By way of example, think of the straight lines of Main Street, railway bridges, and the floodway in Winnipeg, alternately bypassing, crossing, and corralling the serpentine Red and Assiniboine Rivers. First and second natures are layered but not stratified; the historical constantly dips into the ecological for resources and inspiration, and leaves a foot-print that cannot be erased. Historic sites record how we encountered the natural world – with what plans, ideas, technologies – and how we have transformed it. They are our field notes.

We have not always recognized them as such. National historic sites have been tasked with telling (versions of) the story of nation build-ing as collective achievement, as making a homeland and generating prosperity and well-being. There are certainly numerous ways to chal-lenge and diversify that mandate, especially as others are doing in re-centring the settler relationship with Indigenous lands and peoples. Another approach is to read nation building through an environ-mental lens. These sites can't physically reconstruct a historical ecosys-tem; we can't re-submerge the shoreline at L'Anse aux Meadows or replant millions of acres of native grasses. But we can examine why these moments of "human creativity" happened in these places, and what happened as a result. As historian John Wadland observed so in-sightfully, "There is a vast difference between knowing based on what is brought to the place and knowing that emerges from it."[3] There is an enormous amount of environmental knowledge contained in these places: stories of hubris, adaptation, failure, exchange, resilience. Be-cause *this* is our history, too. And it is a history that spills out from behind palisades and sod mounds into the choices we must make now.

What does an environmental reading of historic places tell us? First, that we cannot treat designated historic sites as "magic kingdoms," as islands of time *or* space.[4] If public history prefers to assign a site a primary period of significance, environmental history insists it must be read in a larger context. Consider the numerous agricultural tra-ditions that have shaped Grand Pré. It is precisely *because* of those competing traditions that the site can speak to both industrial infra-structures and alternative practices of wetland agriculture, to the regional economy and the local shoreline. Similarly, immersing ourselves in a cowboy fantasy at Pekisko Creek is a delightful way to spend a summer day, but seeing oil derricks over the next hill suddenly exposes the threads of frontier thinking – the arc of "hope and ruin" in industrial expansion – tying the Bar U to the tar sands and beyond.[5] We need to be willing to think big, to look for national stories in environmental change as well as nation building.

Second, these sites can be read as analogies to some of our current dilemmas. Studying history doesn't enable us to predict the future, but it does let us identify patterns and make comparisons. At L'Anse aux Meadows and other sites on the North Atlantic, we can draw parallels between eras of climate change, scales of harvest, and types of fuel. This puts our current practices in an environmental story of succession, and in an awareness of limits that the heroic story of nation building does not imagine. As one of the early interpretative plans for L'Anse suggested, this site shows how different groups have shared "a basic dependence for survival on the same set of natural resources."[6] At the same time, revisiting the fate of the medieval Norse may give us pause about our own seagoing confidence in claiming the vast waters and near-endless shorelines (and sub-seabeds) of the subarctic north. By offering precedents and points of comparison, we can consider the longer view of sustainability in a post-industrial world.

Finally, historic sites explain the making of a Canadian habitat. Here I want to emphasize both the Canadian – in the making of a national imagination, its preferences and prejudices – and the habitat, the transformative effect of that imagination on the land it claimed as territory. What kinds of nature has been valued, and by whom, and why? Who has had access, and for what purpose? What was attractive about a "wilderness" view at the reconstructed Fort William? How have we valorized an extractive resource economy in romantic stories of frontier expanse? Where do we continue to do so? Historic sites are not "a break from the every day," as Parks Canada says on its website, but a mirror *to* our every day. If we think of historic sites as samples of our habitat rather than designated exceptions from another time, we may see our own motivations more clearly. History is not meant to be merely an affirmation of who we are or have become. Environmental history, especially, asks us to acknowledge that our choices have consequences, and costs.

So where to from here? We certainly need more research about historic sites – comparative, contextual work that locates them in a wider
political and ecological trajectory. What does it say that we have one
monograph, now a quarter of a century old, on the subject of historic
sites? As "national treasures" they are carefully shielded from critique, even when the health and direction of Parks Canada warrants
serious attention and discussion. If former Parks Canada Agency historians talk about the 1970s as a golden age for historical research,
what does that say about the state of affairs today?[7] Why has there
been an erosion of intellectual resources and, it has to be said, intellectual freedom within the agency for historical work? When we see
site plans authorizing historic sites to cultivate environmental citizenship and stewardship replaced with marketing for adventure or a
scenic wedding venue, what are we to make of this? What does this
say about our political priorities, especially when supposedly we are
concerned about environmental sustainability?

For most of the twentieth century, Canada had a category of "National Historic Parks." (L'Anse aux Meadows was one such park.)
These were to be historic properties "of suitable size," big enough to
include nature in the story of the historic event. The concept has faded
in favour of parks *and* historic sites, but how would our history look
if we reintegrated them?[8] Each of the five sites explored here has the
scale, ecological diversity, and environmental history to be presented
as such a park. Each could re-emphasize Canada as a settler project,
with all its social and environmental implications. Each could speak
the relationship *between* ecological and commemorative integrity, and
finally acknowledge our role in "human interaction with the environment especially when it has become vulnerable under the impact
of irreversible change," as UNESCO requires. Heritage designation can
and should require a frank discussion about sustainable and unsustainable practices in history.

We must use the knowledge contained in our historic landscapes in
the ever-more critical planning for sustainability. This means connecting one element of civic consciousness to another: connecting
national history to national direction in issues like climate change and
resource extraction; asking what is often a nostalgic and recreational
experience of history to engage with harder questions of collective

action and common values. Too often environmental policy is framed in terms of present (the crisis at hand) and future (inevitable degradation, or redemptive solution). These historic places embody the value of environmental *history* in today's world. They offer diagnostic insight into the origins, extent, and longevity of the problem. They humanize the story – often told through data and scientific terms – as one of people and choices. They record the unintended consequences of those choices. They may suggest alternatives. They demarcate limits as well as potential. And they ask us to use our imaginations about the past, and for the future.

Notes

INTRODUCTION

1 I owe the phrase "braided narrative" of history to James Rice, from the 2010 Arpents de Neige meeting at Kingston, Ontario. For the introduction to *The Sky on Location* and the epigraph of this chapter, I thank Rebecca Meyers of Bucknell University's Film and Media Studies Program.

2 James Reaney, "Winnipeg Seen as a Body of Time and Space," first published in *A Message to Winnipeg* (1962) and then in *Selected Longer Poems* (Erin, ON: Press Porcépic, 1976).

3 Graeme Wynn and Matthew Evenden, "54, 40 or Fight: Writing within and across Borders in North American Environmental History," in *Nature's End: History and Environment*, ed. Sverker Sörlin and Paul Warde (London: Palgrave, 2009), 230.

4 Cole Harris, *The Reluctant Land: Society, Space, and Environment in Canada before Confederation* (Vancouver: University of British Columbia Press, 2008); Graeme Wynn, *Canada and Arctic North America: An Environmental History* (Santa Barbara, CA: ABC-CLIO, 2006); Neil Forkey, *Canadians and the Natural Environment to the Twenty-First Century* (Toronto: University of Toronto Press, 2012); Laurel Sefton MacDowell, *An Environmental History of Canada* (Vancouver: University of British Columbia Press, 2012).

5 Parks Canada, *National Historic Sites System Plan* (Ottawa: Her Majesty the Queen in Right of Canada, 1997, 2000), http://www.pc.gc.ca/docs/r/system-reseau/sites-lieux1.aspx.

6 Thomas H.B. Symons, "Commemorating Canada's Past: From Old Crow to New Bergthal," in *The Place of History: Commemorating Canada's Past* (Proceedings of the National Symposium held on the Occasion of the 75th Anniversary of the Historic Sites and Monuments Board), ed. T.H.B. Symons (Ottawa: Royal Society of Canada, 1997), 14.

7 Canada National Parks Act, S.C. 2000, c. 32 (Section 4), although the language dates to the National Parks Act of 1930. See Claire E. Campbell, ed., *A Century of Parks Canada, 1911–2011* (Calgary: University of Calgary Press, 2011). Canada was the first country in the world to create a national agency dedicated to managing national parks in 1911, five years ahead of the United States. But the phrase "unimpaired for future generations" had appeared fourteen years earlier in the legislation founding the American National Parks Service.

8 UNESCO, Convention Concerning the Protection of the World Cultural and Natural Heritage, 16 November 1972, http://whc.unesco.org/en/conventiontext/. Subsequent conventions broadened the definition of heritage to include historic landscapes, while other International Council on Monuments and Sites (ICOMOS) charters have expressed environmental concerns such as pollution and overvisitation at historic sites. There are some "mixed sites" but most fall into the categories of cultural or natural. We have begun to see terms like *biocultural diversity* and *integrated heritage conservation*; see David Harmon, "A Bridge over the Chasm: Finding Ways to Achieve Integrated Natural and Cultural Heritage Conservation," *International Journal of Heritage Studies* 13, no. 4–5 (2007): 380–92.

    Since the UNESCO convention recognizes state signatories, Parks Canada is responsible for the formal nomination and administration of World Heritage Sites in Canada.

9 Peter Seixas, "What Is Historical Consciousness?" in *To the Past: History Education, Public Memory, and Citizenship in Canada*, ed. Ruth Sandwell (Toronto: University of Toronto Press, 2006), 11.

10 An example that considers "place" from a perspective of public, rather environmental, history perspective is James Opp and John Walsh, eds., *Placing Memory and Remembering Place in Canada*

(Vancouver: University of British Columbia Press, 2010). C.J. Taylor's *Negotiating the Past: The Making of Canada's National Historic Parks and Sites* (Montreal and Kingston: McGill-Queen's University Press, 1990) remains the only book on the subject. Other useful material includes a special issue of *The Public Historian* on the National Park Service and historic preservation (9, no. 2, 1987); Shannon Ricketts, "Cultural Selection and National Identity: Establishing Historic Sites in a National Framework, 1920–1939," *The Public Historian* 18, no. 3 (1996), 23–41; David Harmon, Francis MacManamon, and Dwight Pitcaithley, eds., *The Antiquities Act: A Century of American Archaeology, Historic Preservation, and Nature Conversation* (Tucson: University of Arizona Press, 2006); Claire Campbell, "'It Was Canadian, Then, Typically Canadian': Revisiting Wilderness at Historic Sites," *British Journal of Canadian Studies* 21, no.1 (2008): 5–34 and "On Fertile Ground: Locating Historic Sites in the Landscapes of Fundy and the Foothills," *Journal of the Canadian Historical Association/Revue de la Société Historique du Canada* 17, no. 1 (2006): 235–65; Cecilia L. Morgan, *Commemorating Canada: History, Heritage, and Memory, 1850s–1990s* (Toronto: University of Toronto Press, 2016). Larry Ostola, former vice-president with Parks Canada, summarizes the institutional history in "Parks Canada's National Historic Sites: Past, Present, and Future," *The George Wright Forum*, 27, no. 2 (2010): 161–69.

A good example of the writing of history from within national park boundaries is I.S. MacLaren, ed., *Culturing Wilderness in Jasper National Park: Studies in Two Centuries of Human History in the Upper Athabasca River Watershed* (Edmonton: University of Alberta Press, 2007).

Rare examples of the intersection between public and environmental history are: two special issues of *The Public Historian*, 26, no. 1 (2004) and 36, no. 3 (2014); also Martin V. Melosi and Philip Scarpino, eds., *Public History and the Environment* (Malabar, FL: Krieger Publishing, 2004), especially the essay by Carl Shull and Dwight T. Pitcaithley, "Melding the Environment and Public History: The Evolution and Maturation of the National Park Service," 56–71.

11 M.B. Payne and C.J. Taylor, "Western Canadian Fur Trade Sites and the Iconography of Public Memory," *Manitoba History* 46 (2003–04): 2–14. While this book deals with national sites, Quebec showed an early interest in historic preservation (because of particular concerns over "national culture" and "national property"), including a 1935 act focused on preserving the historic and rural landscape of Île d'Orléans.

12 See Taylor, *Negotiating the Past*, 15–17 and 28–9. As Max Page and Randall Mason note, historic preservation emerged as "part of a broad effort among Progressive reformers to transform the nature of urban space – its aesthetic character, its social uses, what it signified to society, how it was used, and who controlled it – as a means of transforming society." *Giving Preservation a History: Histories of Historic Preservation in the United States* (New York: Routledge, 2014), 11. Environmentalism, of course, has its roots in the Progressive era as well.

13 There is a vast literature on public history in the United States, and a much more substantial body of work on the history of the National Park Service as well. A useful starting point is Denise Meringolo, *Museums, Monuments, and National Parks: Toward a New Genealogy of Public History* (Amherst: University of Massachusetts, 2012).

14 Alan Gordon, *Time Travel: Tourism and the Rise of the Living History Museum in Mid-Twentieth-Century Canada* (Vancouver: University of British Columbia Press, 2016).

15 Royal Commission on National Development in the Arts, Letters and Sciences, *Report* (Ottawa: King's Printer, 1951), 123. The Historic Sites and Monuments Act (1952–53) outlines the ministerial powers regarding the commemoration of historic places and the role of the Historic Sites and Monuments Board of Canada.

16 In the late 1980s, the Canadian Museum of Civilization took a similar approach, trying to recreate "evocative environments" in three-dimensional spaces to convey "the sense of spirit and adventure that moves Canadians from one frontier to another." Peter E. Rider, "Presenting the Public's History to the Public: The Case of the Canadian Museum of Civilization," in *Studies in History and*

Museums, ed. Peter E. Rider (Hull, QC: Canadian Museum of Civilization, 1994), 87–8.

17 Cole Harris, "The Emotional Structure of Canadian Regionalism," The Walter L. Gordon Lecture Series 1980–81, vol. 5, *The Challenges of Canada's Regional Diversity* (Toronto: Canada Studies Foundation, 1981), 9–30. For the original articulation of limited identities in the postwar period, see William Morton, "Clio in Canada: The Interpretation of Canadian History," *University of Toronto Quarterly* 15, no. 3 (1946): 27–34; Ramsay Cook, "Canadian Centennial Celebrations," *International Journal* 22 (1967): 659–63; J.M.S. Careless, "Limited Identities in Canada," *Canadian Historical Review* 50, no. 1 (1969): 1–10.

18 Louis Applebaum and Jacques Hébert, *Report of the Federal Cultural Policy Review Committee* (Ottawa: Information Services, Dept. of Communications, Government of Canada, 1982), 118; Bernard Ostry, *The Cultural Connection: An Essay on Culture and Government Policy in Canada* (Toronto: McClelland and Stewart, 1978). In 1972, Gérard Pelletier, the secretary of state, had announced a new National Museums Policy, based on the twin principles of "democratization and decentralization." While much of the national attention in this era was focused on reconciling French and English Canada, the "new west" crafted a heritage framework that mirrored its growing influence in and attitude toward the political theatre and national economy. See, for example, Mark Rasmussen, "The Heritage Boom: The Evolution of Historical Resource Conservation in Alberta," *Prairie Forum* 15, no. 2 (1990): 235–62; Patricia K. Wood, "The Historic Site as Cultural Text: A Geography of Heritage in Calgary, Alberta," *Material History Review* 52 (2000): 33–43; James Opp, "Prairie Commemorations and the Nation: The Golden Jubilees of Alberta and Saskatchewan, 1955," in *Canada of the Mind: The Making and Unmaking of Canadian Nationalisms in the Twentieth Century*, ed. N. Hillmer and A. Chapnick (Montreal and Kingston: McGill-Queen's University Press, 2007), 214–33.

19 Parks Canada defines cultural landscape "as any geographical area that has been modified, influenced or given special cultural meaning

by people." *Standards and Guidelines for the Conservation of Historic Places in Canada*, 2nd ed. (Ottawa: Her Majesty the Queen in Right of Canada, 2010), 49. Cultural landscapes are frequently grouped into three categories: designed (or formal), vernacular, and associative; for definitions see UNESCO, *Operational Guidelines for the Implementation of the World Heritage Convention* (Paris: World Heritage Centre, 2008), 86.

A good overview of the development of cultural landscape thinking can be found in Susan Buggey, *An Approach to Aboriginal Cultural Landscapes* (Ottawa: Historic Sites and Monuments Board of Canada, 1999), and Buggey and Nora Mitchell, "Cultural Landscapes: Venue for Community-Based Conservation," in *Cultural Landscapes: Balancing Nature and Heritage in Preservation Practice*, ed. Richard Longstreth (Minneapolis: University of Minnesota Press, 2008). Without the same context and motivation of aboriginal land rights, the American National Park Service came a bit later to the concept of cultural landscape; it was prompted more by questions of how to designate and manage vernacular and associated landscapes in rural areas that had established clusters of buildings and patterns of land-use. See Robert Melnick, *Cultural Landscapes: Rural Historic Districts in the National Park System* (National Park Service, 1984) and his later work, including *Preserving cultural landscapes in America*, edited with Arnold R. Alanen (Baltimore: Johns Hopkins University Press, 2000); also Melody Webb, "Cultural Landscapes in the National Park Service," *The Public Historian* 9, no. 2 (1987): 77–89. See also Richard White, "From Wilderness to Hybrid Landscapes: The Cultural Turn in Environmental History," *The Historian* 66 (2004), 558. The "cultural turn" is apparent in the park history that has proliferated in Canada in recent years.

20 Both Roosevelt administrations expanded federal lands, protective designations, and conservation projects; see, for example, Douglas Brinkley's *The Wilderness Warrior: Theodore Roosevelt and the Crusade for America* (New York: Harper Collins, 2009) and *Rightful Heritage: Franklin D. Roosevelt and the Land of America* (New York: Harper, 2016). For an overview of the pendulum swinging between development, conservation (or "wise use"), and protection

in the national parks, see Richard West Sellars, *Preserving Nature in the National Parks: A History* (New Haven, CT: Yale University Press, 1997) and Ethan Carr, *Mission 66: Modernism and the National Park Dilemma* (Amherst: University of Massachusetts Press in association with Library of American Landscape History, 2007). In the Canadian case, an excellent glimpse of the debates over park use or protection by mid-century are the compilations J.G. Nelson and R.C. Scace, eds., *The Canadian National Parks: Today and Tomorrow* (Calgary: National and Provincial Parks Association of Canada and the University of Calgary, 1969) and J.G. Nelson with R.C. Scace, eds., *Canadian Parks in Perspective: Based on the conference The Canadian National Parks: Today and Romorrow, Calgary, October 1968* (Montreal: Harvest House, 1970).

Late-century environmentalism appears to be a new area of growth in environmental history; see Frank Zelko, *Make It a Green Peace: The Rise of Countercultural Environmentalism* (New York: Oxford University Press, 2013) and several of the monographs in the Nature/History/Society Series with the University of British Columbia Press, including Justin Page, *Tracking the Great Bear: How Environmentalists Recreated British Columbia's Coastal Rainforest* (2014); Ryan O'Connor, *The First Green Wave: Pollution Probe and the Origins of Environmental Activism in Ontario* (2015); and Mark Leeming, *In Defence of Home Places: Environmental Activism in Nova Scotia* (2017).

That said, in protecting both natural and cultural heritage, Canada generally comes later to the table than our neighbour to the south. The United States established a registry of national landmarks (1960), a national wilderness act (1964), and a national historic preservation act (1966). Then, the National Park Service attempted to inject some environmental messaging into site interpretation in the 1970s, in response to the political climate of second-wave environmentalism, but with some resistance from historians. See Barry Mackintosh, *Interpretation in the National Parks Service: A Historical Perspective* (Washington, DC: National Park Service, 1986), 69. While there was also great momentum in Canada in the late 1960s toward museums, heritage, and national park creation, it is worth noting Canada never created a wilderness act, and the Canadian Register of Historic Places only came into existence in 2001.

21 Parks Canada, *Guiding Principles and Operational Policies* (Ottawa: Minister of Supply and Services Canada, 1994). We certainly saw fewer references to national historic sites during the Parks Canada centennial than to "glamour shots" of scenic (and unpeopled) beauty in national parks.

22 Parks Canada, *Fort Beauséjour and Fort Gaspareaux National Historic Sites Management Plan* (Hull, QC: Parks Canada, 1997), 31.

23 Henry Commager, "The Search for a Usable Past," originally published in *American Heritage* 16, no. 2 (February 1965), reprinted in *The Search for a Usable Past and Other Essays in Historiography* (New York: Knopf, 1967), 21.

24 John E. Bodnar, *Remaking America: Public Memory, Commemoration, and Patriotism in the Twentieth Century* (Princeton: Princeton University Press, 1992), 177.

25 See M. Brook Taylor, *Promoters, Patriots, and Partisans: Historiography in Nineteenth-Century English Canada* (Toronto: University of Toronto Press, 1989).

26 As Audrey J. Horning observed of a "founding" site in the United States,"Unremitting emphasis upon the proto-American nature of Jamestown has threatened our ability to understand the actualities of its seventeenth-century past, the daily experiences of its occupants and visitors, who were far more English or Irish or Dutch or African in outlook than America, let alone Southern American." "Of Saints and Sinners: Mythic Landscapes of the Old and New South," in *Myth, Memory, and the Making of the American Landscape*, ed. Paul Shackel (Gainsville: University Press of Florida, 2001), 24.

27 Benedict Anderson, *Imagined Communities: Reflections on the Origin and Spread of Nationalism*, 2nd ed. (London: Verso, 1991), 182. This is part of his argument as to how the census, the museum, and the map enabled the colonial state "profoundly shaped the way in which the colonial state imagine its dominion – the nature of the human beings it ruled, the geography of its domain, and the legitimacy of its ancestry," 164. My categories here align with

George Altmeyer, "Three Ideas of Nature in Canada, 1893–1914," *Journal of Canadian Studies* 11, no. 3 (1976): 21–36.

28 A.B. McKillop, *A Disciplined Intelligence: Critical Inquiry and Canadian Thought in the Victorian Era* (Montreal and Kingston: McGill-Queen's University Press, 2001), 231. As he says elsewhere, this suggests an earlier golden age of ideas and influence; McKillop, "Public Intellectuals and Canadian Intellectual History: Communities of Concern," in *Les idées en mouvement: Perspectives en histoire intellectuelle et culturelle du Canada*, ed. Damien-Claude Bélanger, Sophie Coupal, and Michel Ducharme (Quebec: Presses de l'Université Laval, 2004). The most famous and influential voice of this conservative resurgence was that of Jack Granatstein's *Who Killed Canadian History?* (Toronto: Harper Collins, 1998).

29 Frits Pannekoek, "Who Matters?: Public History and the Invention of the Canadian Past," *Acadiensis* 24, no. 2 (2000) argues that historic sites sustain a most conservative version of history, although he notes that academic historians serving on the HSMBC who have in fact shaped public commemoration; David Lowenthal, *The Heritage Crusade and the Spoils of History* (Cambridge: Cambridge University Press, 1998).

30 Nicole Neatby and Peter Hodgins, eds., *Settling and Unsettling Memories: Essays in Canadian Public History* (Toronto: University of Toronto Press, 2012), 6.

31 The extreme reaction to *The West as America: Reinterpreting Images of the Frontier, 1820–1920*, the National Museum of American Art's 1991 exhibit of frontier art, at the height of the "culture wars" between nationalist and revisionist historians, indicated that an academic revision of mythic images can provoke extreme sentiments, given that many remain deeply attached to conventional, romantic, and nationalist ideas of the historical landscape. Some argued that the exhibit's unpopularity was due to the public preference for nostalgic, uncontroversial, untroubling pasts; others, that the fault lay with the heavy-handed academic curatorial language, which "neglected the public audience" and failed to respect their "defining cultural myths." See Thomas A. Woods, "Museums and the Public: Doing History Together," *Journal of American History* 82, no. 3 (1995): 1114; also Roger B. Stein, "Visualizing Conflict

in the West as America," *The Public Historian* 14, no. 3 (1992): 85–91; and Steven C. Dubin, *Displays of Power: Memory and Amnesia in the American Museum* (New York: New York University Press, 1999), 152–85.

On the other hand, Canadian curators Glenn C. Sutter and Douglas Worts argue that as generally respected and trusted public institutions, museums have a social responsibility to teach about sustainability: "Our goal should be to reinforce, embellish, or alter what people know about their world (their ecological literacy) and how they relate to it (their ecological identity)" in "Negotiating a Sustainable Path: Museums and Societal Therapy," in *Looking Reality in the Eye: Museums and Social Responsibility*, eds. Robert R. Janes and Gerald T. Conaty (Calgary: University of Calgary Press, 2005), 143.

32 B.A. (Sandy) Balcom, "Reflections on Authenticity at a Reconstructed Site, 1976–1979, 1987–2012," in Forum: Louisbourg Researchers Recall their Roles in the Reconstruction of Louisbourg, 1961–2013, *The Nashwaak Review* 30–1, no. 1 (2013): 423.

33 Anne Marie Jonah and Chantal Vechembre, *French Taste in Atlantic Canada, 1604–1758* (Sydney: Cape Breton University Press, 2012); A.J.B. Johnston, *Louisbourg: Past, Present, Future* (Halifax: Nimbus, 2013). Consider also the work of former Parks Canada historians Lyle Dick and Simon Evans in researching historic sites in the west and north. Nor is this is limited to environmental issues; Laura Peer, for example, has argued for using historic sites to educate and change public opinion about Indigenous histories and status. *Playing Ourselves: Interpreting Native Histories at Historic Reconstructions* (Lanham, MD: AltaMira Press, 2007).

34 Meringolo, *Museums, Monuments*, xvii.

35 Carol Galbreath, "Historic Preservation as an Environmental Problem," *California Historian* 21, no. 2 (1974): 6. Active history is defined "variously as history that listens and is responsive; history that will make a tangible difference in people's lives; history that makes an intervention and is transformative to both practitioners and communities. We seek a practice of history that emphasizes collegiality, builds community among active historians and other members of communities, and recognizes the public responsibilities

of the historian." "Active History: About," accessed 20 February 2017, http://activehistory.ca/about/.

One of the most oft-quoted dictums about public history comes from a 1957 book by Freeman Tilden, an interpreter with the US National Park Service: "The chief aim of interpretation is not instruction but provocation." I would suggest engagement, rather than provocation: to refuse to detach the past from the present, history from citizenship. Freeman Tilden, *Interpreting Our Heritage*, rev. ed. (Chapel Hill: University of North Carolina Press, 1967), 9.

36 Parks Canada, *Standards and Guidelines*, 3; Melody Webb, "Cultural Landscapes in the National Park Service," *The Public Historian* 9, no. 2 (1987): 77–89; Bill Freedman, Michael Macdonald, and Harry Beach, *Ecological Conditions at the Grand-Pré National Historic Site* (Halifax: Parks Canada, 2001).

37 Robert Z. Melnick, "Are We There Yet? Travels and Tribulations in the Cultural Landscape," in *Cultural Landscapes: Balancing Nature and Heritage in Preservation Practice*, ed. Richard Longstreth (Minneapolis: University of Minnesota Press, 2008), 201; Rebecca Conard, "Applied Environmentalism, or Reconciliation Among 'the Bios' and 'the Culturals,'" *The Public Historian* 23, no. 2 (2001): 17.

38 Laura A. Watt, Leigh Raymond, and Meryl L. Eschen, "Reflections on Preserving Ecological and Cultural Landscapes," *Environmental History* 9, no. 4 (2004): 620–47.

39 Parks Canada, *Standards and Guidelines*, 67.

40 Parks Canada, *Principles and Guidelines for Ecological Restoration in Canada's Protected Natural Areas* (Ottawa: Parks Canada and the Canadian Parks Council, 2008).

41 A view from within Parks Canada is Stephen J. Malins, "Convergence and Collaboration: Integrating Cultural and Natural Resource Management" (master's thesis, Royal Roads University, 2011).

CHAPTER ONE

1 For an excellent physical description of the site, see Birgitta Wallace, "L'Anse aux Meadows and Vinland: An Abandoned Experiment," in *Contact, Continuity, and Collapse: The Norse*

Colonization of the North Atlantic, ed. James H. Barrett (Turnhout, Belgium: Brepols, 2003), 207–38; "Viking Village Declared World Site," *Globe and Mail*, 12 July 1980. Helge and Anne Ingstad translated the name literally the Bay of the Meadows, in *The Viking Discovery of America: The Excavation of a Norse Settlement in L'Anse aux Meadows, Newfoundland* (St John's: Breakwater Books, 2000), 133. Parks Canada and Birgitta Wallace, however, read it as an anglicized variant of *L'Anse à la Médée,* or Médée Bay. See Wallace, "Situating L'Anse Aux Meadows in the Vinland Sagas," unpublished paper, 2014, http://www.inl.is/doc/2805.

2 H.J. Porter and Associates Ltd., "Preliminary Development Concept: L'Anse aux Meadows National Historic Park" (Halifax: Parks Canada, Indian and Northern Affairs, 1975), 1–2.

3 Jane Harrison, "Mounds, Middens, and Social Landscapes: Viking-Norse Settlement of the North Atlantic, c. AD 850–1250," in *Northscapes: History, Technology, and the Making of Northern Environments*, ed. Dolly Jørgensen and Sverker Sörlin (Vancouver: University of British Columbia Press, 2013), 85.

4 Robert Grant Halliburton, "'The Men of the North and Their Place in History,' A Lecture Delivered Before the Montreal Literary Club, March 31st, 1869" (Montreal: John Lowell, 1869), 10, emphasis in the original. See also Carl Berger, *The Sense of Power: Studies in the Ideas of Canadian Imperialism* (Toronto: University of Toronto Press, 1970); Renee Hulan, *Northern Experience and the Myths of Canadian Culture* (Montreal and Kingston: McGill-Queen's University Press, 2002); Megan Arnott, "Putting the Vikings on the Canadian Map," in *Mapping Medievalism at the Canadian Frontier*, ed. Katheryn Brush (London: Museum London/Macintosh Gallery, 2010), 108–23.

5 Ian McKay and Robin Bates, *In the Province of History: The Making of the Public Past in Twentieth-Century Nova Scotia* (Montreal and Kingston: McGill-Queen's University Press), 317–27. On the Yarmouth Runic Stone, which was dismissed by local historians even as it was touted by local promoters, see Moses H. Nickerson, "A Short Note on the Yarmouth 'Runic Stone" and Harry Piers, "Remarks on the Fletcher and Related Stones of Yarmouth, N.S.,"

Collections of the Royal Nova Scotia Historical Society 17 (1913):
51–6; Brian Cuthbertson, "Voyages to North America before John
Cabot: Separating Fact from Fiction," *Collections* 44 (1995):
121–44; Mats G. Larsson, "The Vinland Sagas and Nova Scotia:
A Reappraisal of an Old Theory," *Scandinavian Studies* 64, no. 3
(1992): 305–34.

6 See Magnus Magnusson and Hermann Pálsson, eds., *The Vinland
Sagas: The Norse Discovery of America* (London: Penguin Classics,
1965). Also W.A. Munn, *Wineland Voyages: Location of Hellu-
land, Markland, and Vinland* (St John's: The Evening Telegram
Ltd., 1914); Andrew Wawn, "Victorian Vinland," in *Approaches
to Vinland: A Conference on the Written and Archaeological
Sources for the Norse Settlements in the North Atlantic*, ed. Wawn
Andrew and Þórunn Sigurðardóttir (Reykjavík: Sigurður Nordal
Institute, 2001); Arnott, "Putting the Vikings on the Canadian
Map."

7 Helge Ingstad, "Where Was Vinland?," *Land under the Pole Star*,
trans. Naomi Walford (n.p.: St Martin's Press, 1966), 153–66, 171;
Anne Stine Ingstad and Helge Ingstad, *The Norse Discovery of
America*, vol. 2, *Historical Aspects and Other Background Matter*
(Oslo: Norwegian University Press, 1985).

8 "There appears to be no evidence to definitely contradict Dr. Ingstad's identification of the site and the consensus of those who
have visited the excavation and are in a position to judge seems to
be that it is indisputably Norse." John H. Rick, "L'Anse aux Meadows site, Newfoundland," Agenda paper 68-40, Historic Sites and
Monuments Board of Canada, 1968.

9 World Heritage Site entry memo from Ernest Allen Connally, Secretary General, International Council on Monuments and Sites to
Firouz Bagerzadeh, Chairman of World Heritage Committee,
United Nations Education, Scientific, and Cultural Organization
(UNESCO), 7 June 1978; UNESCO Advisory Body Evaluation, 7
April 1978, http://whc.unesco.org/en/list/4. L'Anse was formally
unveiled as a World Heritage Site in July 1980. Meanwhile, archaeologists working in the Far North were finding evidence of Norse
contact and trade with Thule Inuit; see note 45.

10 Malcolm Grey, "Thousands Go to Newfoundland to See Where

Vikings Once Lived," *Globe and Mail*, 26 March 1979. See Kevin McAleese, "L'Anse aux Meadows: Rediscovered and Remade," 181–8, and Darrell Markewitz, "The 'Viking Encampment' at L'Anse aux Meadows National Historic Site of Canada: Presenting the Past," 193–202, in *Vinland Revisited: The Norse World at the Turn of the First Millennium*, ed. Shannon Lewis-Simpson (St John's: Historic Sites Association of Newfoundland and Labrador, 2003). Visitation numbers by Parks Canada at http://www.pc.gc.ca/docs/pc/attend/table2.aspx.

11 Birgitta Wallace, "The Later Excavations at L'Anse aux Meadows," in *Vinland Revisited*, 165–80.

12 Ingstad and Ingstad, *The Viking Discovery of America*, 143.

13 David Gidmark, "Newfie Village Predates Columbus by 500 Years," *Montreal Gazette*, 17 January 1976; Charles Lindsay, "Was L'Anse aux Meadows a Norse Outpost?," *Canadian Geographical Journal* 94, no. 1 (1977): 36–43; Marcia Douglas, "A Voyage in Time to the Viking Age," *Globe and Mail*, 4 May 1985.

14 Gray, "Thousands Go to Newfoundland," *Globe and Mail*, 26 March 1979.

15 "Where Is Vinland?," *Great Unsolved Mysteries in Canada*, accessed 8 February 2017, http://www.canadianmysteries.ca/sites/vinland/indexen.html.

16 David Blackwood, *L'Anse aux Meadows, Newfoundland* (etching), 1985, accessed 8 February 2017, http://www.mayberryfineart.com/artwork/AW1992.

17 "Vikings," Heritage Minutes, Historica Canada, released 1992, access 8 February 2017, https://www.historicacanada.ca/content/heritage-minutes/vikings.

18 Bengt Schonback and Nicolas Dykes, "L'Anse aux Meadows National Historic Park: Interpretive Plan and Exhibit Storyline," first draft, 22 April 1974, Ottawa, as Appendix A in *L'Anse aux Meadows National Historic Park Interpretation Programme: Preliminary VRC Interpretation Concept and Interpretative User Requirements* (Halifax: Parks Canada, 1980), 7 and 11; H.J. Porter and Associates, "Preliminary Development Concept," 9.

19 Parks Canada, *L'Anse aux Meadows National Historic Site* brochure, 2014; Parks Canada, *L'Anse aux Meadows National*

Historic Site of Canada Management Plan (Ottawa: Parks Canada, 2003), 8, 24. The management plan also notes that only about half visitors actually go out to site (25). The hiking trail is a remnant of the site's original designation as a "national historic park" – original plans had several such trails.

20 See Claire Campbell, "On Fertile Ground: Locating Historic Sites in the Landscapes of Fundy and the Foothills," *Journal of the Canadian Historical Association* 17, no. 1 (2006): 235–65. Good examples include Head-Smashed-In Buffalo Jump in Alberta, where the external architecture aligns with the geological layers in the hillside while the exhibits inside follow the course of the jump. Another example of sympathetic architecture can be found at Port au Choix, where the visitor centre borrows both from Inuit structures and local fishing sheds. One proposal for L'Anse suggested a centre based not on the landscape but (as with Port au Choix) on the historical occupation, segmenting the exhibits in a cluster of "Norse-style" buildings ("Preliminary VRC Interpretative Concept," 15), but the 1975 plan felt the architecture that echoed "the rugged rock outcrop" and avoided any reference to Norse design would be more authentic (49), and it was this that won out. Ironically, though, the Visitor Reception Centre (VRC) was stalled for ten years because it had to be redesigned on a smaller scale and budget.

21 In "The 'Viking Encampment' at L'Anse aux Meadows National Historic Site of Canada," Markewitz decries what he sees as the watering down of the authenticity of the site in favour of a more typical living history. The Nara Document on Authenticity (ICOMOS, 1994) recognizes that our valuation of authenticity may derive from any number of factors, which "may include form and design, materials and substance, use and function, traditions and techniques, location and setting, and spirit and feeling, and other internal and external factors" (Item 13). My point is not to make light of the intangible factors (spirit, feeling, traditions, etc.) in *determining* authenticity, but to argue that physical objects, something the public can see and touch and walk around, will have a greater impact in the communication *of* that authenticity.

22 Parks Canada, *National Historic Sites System Plan* (Ottawa: Her Majesty the Queen in Right of Canada, 1997, 2000).

23 Philip Gimbarzevsky, *L'Anse aux Meadows National Historic Park Integrated Survey of Biophysical Resources* (Ottawa: Environment Canada, 1977), 1. For more on the history of the national park system and the pedigree of the concept of "unimpaired," see Claire E. Campbell, ed., *A Century of Parks Canada, 1911–2011* (Calgary: University of Calgary Press, 2011).

24 Wallace, "The Norse in Newfoundland" provides an extensive list. Examples include Gimbarzevsky, *L'Anse aux Meadows National Historic Park*; D.R. Grant, *Surficial Geology and Sea-Level Changes, L'Anse aux Meadows Historic Park, Newfoundland* (Ottawa: Geological Survey of Canada, 1975); Tom H. Northcott, *The Land and Sea Mammals of L'Anse aux Meadows National Historic Park, Newfoundland* (Place: Parks Canada/Indian and Northern Affairs, 1976); Frederick C. Pollett, W.J. Meades, and Alexander W. Robertson, *The Classification and Interpretation of the Vegetation Resource within L'Anse aux Meadows National Historic Park* (n.p.: Parks Canada, 1975); R.D. Lamberton and John E. Maunder, *An Avifaunal Survey of L'Anse aux Meadows National Historic Park* (Ottawa: Parks Canada, 1976). On the pollen record, see A.M. Davis, J.H. McAndrews, and B.L. Wallace, "Paleoenvironment and the Archaeological Record at the L'Anse aux Meadows Site, Newfoundland," *Geoarchaeology: An International Journal* 3, no. 1 (1988): 53–64. McAndrews had done a pollen study for Parks Canada in 1978.

25 Parks Canada, *L'Anse aux Meadows National Historic Site Management Plan*, 13. In 1890, Sir Daniel Wilson, then president of the University of Toronto, told the Royal Society of Canada, ""But there is every motive to stimulate us to a careful survey of the coast in search of any probable site of the Vinland of the old Northmen ... A fresh stimulus is thus furnished to our Canadian yachtsmen to combine historical exploration with a summer's coasting trip, and go in search of the lost Vinland." (He himself leaned toward the coast of Nova Scotia.) "The Vinland of the Northmen," read 27 May 1890, *Transactions of the Royal Society of Canada*, Section II (1890), 120–1. The Ingstads avowed that following the sailing directions of the sagas you couldn't help but discover it. (So the

Ingstads told the NFB, and so the curator of the American Museum
of Natural History told the press.)

A good illustration of the environmental orientation in Vinland
scholarship is the collection *Vinland Revisited: The Norse World at
the Turn of the First Millennium* (2003), in which scholars refer to
various environmental conditions, both terrestrial and marine, of
the Norse era to speculate on the location of Vinland, the duration
of the settlement, and the Norse Atlantic more broadly.

26 Parks Canada, *L'Anse aux Meadows National Historic Site Management Plan*, 13; Canadian Register of Historic Places, "Climate
Change and National Historic Sites," accessed 8 February 2017,
http://www.historicplaces.ca/en/pages/6_climate_change-change
ment_climatique.aspx?pid=58648&h=climate%20change. Also,
the *State of Canada's Natural and Historic Places* reports have
noted that erosion "has worsened over time due to climate-related
effects" as in 2011 (see http://www.pc.gc.ca/docs/pc/rpts/elnhc-
scnhp/2011/part-b.aspx). Research about climate change by and
for Parks Canada has focused on national parks; e.g., Daniel Scott
and Roger Suffling, "Climate Change and Canada's National Park
System: A Screening Level Assessment" (Ottawa: Parks Canada,
2000); Daniel Scott, Jay R. Malcolm, and Christopher Lemieux,
"Climate Change and Modelled Biome Representation in Canada's
National Park System: Implications for System Planning and Park
Mandates," *Global Ecology and Biogeography* 11 (2002): 475–84.

A high percentage of World Heritage Sites are located in low-
lying maritime areas (coastal, islands) – not surprisingly, given the
history of human settlement – which means a high percentage of
sites are affected by sea level rise caused by climate change. Few of
these sites, though, will be affected by glacial rebound specifically,
since "cultural sites, and the current population distribution, are
not concentrated near the last glacial ice masses where the rebound
is strongest." L'Anse aux Meadows is third in Canada after SGang
Gwaay and Lunenburg as vulnerable to changes in temperature.
Ben Marzeion and Anders Levermann, "Loss of Cultural World
Heritage and Currently Inhabited Places to Sea-Level Rise," *Environmental Research Letters* 9, no. 3 (2014): 5 and Table 6, http://

iopscience.iop.org/1748-9326/9/3/034001/media/erl491558supp
data.pdf.

27 Philip Garone, "Mission Convergence?: Climate Change and the
    Management of US Public Lands," *Environmental History* 19, no.
    2 (2014): 350.

28 Trevor Bell, Joyce B. Macpherson, M.A.P. Renouf, "'Wish You
    Were Here ...': A Thumbnail Portrait of the Great Northern Penin-
    sula AD 1000," in *Vinland Revisited*, 203–18; Davis, McAndrews,
    and Wallace, "Paleoenvironment and the Archaeological Record,"
    3. See also J. Shaw, P. Gareau, and R.C. Courtney, "Palaeogeogra-
    phy of Atlantic Canada 13–0 kyr," *Quaternary Science Reviews* 21,
    no. 16–17 (2002): 1861–78; J. Shaw, D.J.W. Piper, G.B.J. Fader,
    E.L. King, BJ Todd, T. Bell, M.J. Batterson, and D.G.E. Liverman,
    "A Conceptual Model of the Deglaciation of Atlantic Canada,"
    Quaternary Science Reviews 25, no. 17–18 (2006): 2059–81.
        Archaeologist Peter Pope describes how the process of isostatic
    rebound creates a series of beach terraces revealing progressively
    older finds for archaeologists, allowing them to "step upward and
    therefore backward in time." "Historical Archaeology and the
    Maritime Cultural Landscape of the Atlantic Fishery," in *Method
    and Meaning in Canadian Environmental History,* ed. Alan
    MacEachern and Bill Turkel (Toronto: Nelson, 2008), 36. The re-
    bound seemed to be a recognized phenomenon at the historic site.
    "Everyone's losing their coastline," said one interpreter ("even
    St John's is losing theirs," another said), "but we're gaining ours"
    (on-site conversation, 25 June 2009).

29 Trevor Bell and M.A.P. Renouf, "By Land and Sea: Landscape and
    Marine Environmental Perspectives on Port au Choix Archaeol-
    ogy," in *The Cultural Landscapes of Port au Choix: Precontact
    Hunter-Gatherers of Northwestern Newfoundland,* ed. M.A.
    Priscilla Renouf (New York: Springer, 2011), 21–42; M.A.P. Re-
    nouf and Trevor Bell, "Contraction and Expansion in Newfound-
    land Prehistory, AD 900–1500," in *The Northern World AD
    900–1400: The Dynamics of Climate, Economy, and Politics in
    Hemispheric Perspective,* ed. Herbert Maschner, Owen Mason, and
    Robert McGhee (Salt Lake City: University of Utah Press, 2009),
    263–78.

30 These epochs are roughly dated (with a significant corollary that
these are not uniform): the Dark Age/Early Medieval Cold Period
500–900, Medieval Warm Period 900–1300, Little Ice Age 1300–
1850, Modern Warm Period/Anthropocene 1850–present. Most
of the (extensive) literature on the historical environments of the
Norse Atlantic consider the Greenlandic settlements (as made possible by a warmer north and decline in sea ice, and weakened by
the onset of cooler weather); Vinland/Newfoundland is mentioned
infrequently, and generally taken as an echo of the Greenlandic fate
(i.e., the furthest of the Norse reach, and the quickest to withdraw
in advance of the collapse of the Greenland colonies). For some
introductions to the historical climatology of the North Atlantic,
see, for example, A.E.J. Ogilvie, L.K. Barlow, and A.E. Jennings,
"North Atlantic Climate *c*. AD 1000: Millennial Reflections on the
Viking Discoveries of Iceland, Greenland and North America,"
Weather 55 (2000): 34–45; Gerold Wefer, Wolfgang H. Berger,
Karl-Ernst Behre, and Eystein Jansen, eds., *Climate Development
and History of the North Atlantic Realm* (New York: Springer,
2002), especially Brian Huntley et al., "Holocene Paleoenviron-
mental Changes in North-West Europe: Climactic Implications and
the Human Dimension" (259–98) and Jean M. Grove, "Climate
Change in Northern Europe over the Last Two Thousand Years
and Its Possible Influence on Human Activity" (313–26); Brian
Fagan, *The Great Warming: Climate Change and the Rise and Fall
of Civilizations* (New York: Bloomsbury, 2008); Knud Frydendahl,
"The Summer Climate in the North Atlantic about the Year 1000,"
in *Viking Voyages to North America*, ed. Birthe L. Clausen (Denmark: Kannike Tryk A/S, 1993), 90–4; extensive work by P.C.
Buckland and Thomas H. McGovern on the Norse expansion dur-
ing the Medieval Warm Period, such as Thomas McGovern, Gerald
Bigelow, Thomas Amorosi, and Daniel Russell, "Northern Islands,
Human Error, and Environmental Degradation: A View of Social
and Ecological Change in the Medieval North Atlantic," *Human
Ecology* 16, no. 3 (1988): 225–70. Thomas Haine situates Norse
navigation technologies within the changing climactic context in
"What Did the Viking Discoverers of America Know of the North
Atlantic Environment?" *Weather* 63 (2008): 60–5. William C.

Foster's *Climate and Culture Change in North America, 900 AD–1600 AD* (Austin: University of Texas Press, 2012) deals primarily on the American south and southwest, but shows that the Medieval Warm Period spurred population growth and agricultural development in North America as well.

31 Magnusson and Pálsson, *The Vinland Sagas*, 56, 98 (although *Eirik's Saga* also describes the first winter as "a very severe one," 95).

32 W.F. Butler, *The Great Lone Land: A Narrative of Travel and Adventure in the North-West of America* (London: S. Low, Marston, Low, & Searle, 1872), 139. On the strategic evaluation of the plains in the nineteenth century, see Lawrence Culver, "Seeing Climate through Culture," *Environmental History* 19, no. 2 (2014): 282–364; Alwynne Beaudoin, "What They Saw: The Climactic and Environmental Context for Euro-Canadian Settlement in Alberta," *Prairie Forum* 40, no. 1 (1999): 1–40; Joan M. Schwartz, "More than 'Competent Description of an Intractably Empty Landscape': A Strategy for Critical Engagement with Historical Photographs," *Historical Geography* 31 (2003): 105–30.

   The introduction to the Forum on Climate Change and Environmental History in *Environmental History* (2014) by Mark Carey and Philip Garone notes the relationship between the global rise of capitalism, industrialization, and climate change in the Anthropocene, drawing our attention to the human economic agenda, colonization and the creation of new settler nation-states, and the climactic effect.

33 See, for example, Janice Cavell and Jeff Noakes, *Acts of Occupation: Canada and Arctic Sovereignty, 1918–1925* (Vancouver: University of British Columbia Press, 2010).

34 "The Fog Lifts over Vinland," *Ottawa Citizen*, 11 November 1961. The *Citizen* praised the Norse as "an incredibly ambitious and valiant crowd." See also P. Whitney Lackenbauer and Matthew Farish, "The Cold War on Canadian Soil: Militarizing a Northern Environment," *Environmental History* 12, no. 4 (2007): 920–50.

35 Trudeau went on to say, "We do not doubt for a moment that the rest of the world would find us at fault, and hold us liable, should

we fail to ensure adequate protection of that environment from pollution or artificial deterioration. Canada will not permit this to happen." Canada, *House of Commons Debates*, 24 October 1969.

36 Gimbarzevsky, *L'Anse aux Meadows National Historic Park*, 7, 31.

37 "Operation NANOOK," National Defence and the Canadian Armed Forces, 19 September 2016, http://www.forces.gc.ca/en/operations-canada-north-america-recurring/op-nanook.page?.

38 The 2013 claim was made to the United Nations Commission on the Law of the Continental Shelf. Russia, France, Denmark, and the United States have adjacent and rival claims. See Government of Canada, *Partial Submission of Canada to the Commission on the Limits of the Continental Shelf Regarding its Continental Shelf in the Atlantic Ocean* (Ottawa: Her Majesty the Queen in Right of Canada, 2013), http://www.un.org/depts/los/clcs_new/submissions_files/can70_13/es_can_en.pdf. See also Jacob Verhoef, David Mosher, and Steve Forbes, "Defining Canada's Extended Continental Shelves," *Geoscience Canada* 38, no. 2 (2011): 85–96. Hans Island to the east and Lomonsov Ridge to the west typify this competition over seabed mineral rights.

On public opinion, see EKOS Research Associates, *Rethinking the Top of the World: Arctic Security Public Opinion Survey, Final Report* (Toronto: Munk-Gordon Arctic Security Program: 2011), http://www.ekospolitics.com/articles/2011-01-25ArcticSecurityReport.pdf. The *Globe and Mail* reported the findings on 24 January 2011, in "Canadians Rank Arctic Sovereignty as Top Foreign Policy Priority."

39 Parks Canada, "The Franklin Expedition," 30 November 2016, http://www.pc.gc.ca/eng/culture/franklin/index.aspx. For a useful comment on the politics of the search, with regards to both nation-building and Indigenous history, see Tina Adcock's "Why Should We Care about the Erebus (or Terror)?" *Active History*, 15 September 2014, http://activehistory.ca/2014/09/why-should-we-care-about-the-erebus-or-terror/. Andrea Charron offers a very readable overview of changing ideas of the Northwest Passage – and their utility to the Canadian national narrative – in "Contesting the Northwest Passage: Four Far-North Narratives," in *Border Flows:*

A Century of the Canadian-American Water Relationship, ed. Lynne Heasley and Daniel MacFarlane (Calgary: University of Calgary Press, 2016), 87–109.

40 As Wallace notes, "The exposed location of the settlement, on the open sea of the Strait of Belle Isle, suggests that seafaring was the most important function of the settlement." "The Norse in New-foundland: L'Anse aux Meadows and Vinland," *Newfoundland Studies* 19, no. 1 (2003): 26. In 2007, the federal government announced the construction of a deepwater port at Nanisivik (a project that has since been scaled down); in 2010, it committed a record $3.6 billion to a national shipbuilding project (or "procurement strategy").

41 Marcy Rockman, "New World with a New Sky: Climatic Variabil-ity, Environmental Expectations, and the Historical Period Colo-nization of Eastern North America," *Historical Archaeology* 44, no. 3 (2010): 4–20. This is an interesting discussion about where environmental knowledge comes from and how it affects our abil-ity to adapt to new places. For example, Rockman writes, "envi-ronmental expectations can draw from community experiences more than a hundred years in the past, and may derive from places lived in prior to the most recent departure point" (8). Thus we might consider a chain of knowledge from Norway to Iceland to Greenland.

42 Magnusson and Pálsson, *The Vinland Sagas*, regarding beauty, 55–6, 95; resources, 65, 67, 98; fortune, 67.

43 Magnus Magnusson, "Vinland: The Ultimate Outpost," in *Vinland Revisited*, 91.

44 Wallace, "The Later Excavations at L'Anse aux Meadows," in *Vin-land Revisited*; "L'Anse aux Meadows and Vinland"; "The Norse in Newfoundland: L'Anse aux Meadows and Vinland," *Newfound-land Studies* 19, no. 1 (2003): 5–43; *Westward Vikings: The Saga of L'Anse aux Meadows* (St John's: Historic Sites Association of Newfoundland and Labrador, 2007). Also Oddvar K. Hoidal, "Norsemen and the North American Forests," *Journal of Forest History* 24 (1980): 200–3.

45 Archaeologists debate whether the Norse would have had contact with the Dorset Inuit or the later Thule Inuit, and which group

were the "Skraelings" described in the sagas. See especially the work of Patricia Sutherland, as in "Strands of Culture Contact: Dorset-European Interactions in the Canadian Eastern Arctic," in *Identities and Cultural Contacts in the Arctic*, ed. Martin Appelt, Joel Berglund, and H.C. Gulløv (Copenhagen: National Museum of Denmark and Danish Polar Centre, 2000; reprinted by the Canadian Museum of History), 159–69; "The Question of Contact between Dorset Palaeo-Eskimos and Early Europeans in the Eastern Arctic," in *The Northern World* AD 900–1400: *The Dynamics of Climate, Economy, and Politics in Hemispheric Perspective*, ed. Herbert Maschner, Owen Mason, and Robert McGhee (Salt Lake City: University of Utah Press, 2009), 279–99; with Peter H. Thompson and Patricia A. Hunt, "Evidence of Early Metalworking in Arctic Canada," *Geoarchaeology* 30, no.1 (2015): 74–8. Also Robert W. Park, "Contact between the Norse Vikings and the Dorset Culture in Arctic Canada," *Antiquity* 82, no. 315 (2008): 189–98; and Peter Schledermann and K.M. McCullough, "Inuit-Norse Contact in the Smith Sound Region," in *Contact, Continuity and Collapse*, 183–206.

Ironically, more Canadians probably learned of Sutherland's research when it ended. In 2013, the CBC television series *The Fifth Estate*, in an episode titled "Silence of the Labs," featured the shuttering of her work on Baffin Island the previous year. The show suggested this was part of a larger "anti-science" mentality in Stephen Harper's Conservatives as well as the kind of political interference that resulted in the Canadian Museum of Civilization being rebranded as "the Canadian Museum of History" with presumably a more nationalist narrative. See http://www.cbc.ca/fifth/episodes/2013-2014/the-silence-of-the-labs.

46 Wallace, "An Abandoned Experiment," especially 212–14 and 228–32; also Peter Pope, who characterizes Norse activities as extended networks of resource foraging rather than settlement, in "Did the Vikings Reach North America without Discovering It?: The Greenland Norse and Zuan Caboto in the Strait of Belle Isle," in *Vinland Revisited*, 350. As Christian Keller explains, "The crucial element here is the geographical setting of these Norse peoples, on the Arctic fringe. The position allowed them to harvest Arctic

and sub-Arctic resources, either directly or indirectly, and profit
from the distribution of these goods to the high end of the European luxury market." "Furs, Fish, and Ivory: Medieval Norsemen
at the Arctic Fringe," *Journal of the North Atlantic* 3, no. 1
(2010): 7–8.

47 William Morton, *The Canadian Identity* (Toronto: University of
Toronto Press, 1961), 5, 91.

48 Magnusson and Pálsson, *The Vinland Sagas*, 65–7 and 99–100.
For a discussion of both the climatic pressures on Norse settlements, and the more successful Inuit adaptations to long-term climate change, see Jared Diamond, *Collapse: How Societies Choose
to Fail or Succeed*, rev. ed. (New York: Penguin, 2011); in response, Joel Berglund, "Did the Medieval Norse Society in Greenland Really Fail?" in *Questioning Collapse: Human Resilience,
Ecological Vulnerability, and the Aftermath of Empire*, ed. Patricia
A. McAnany, and Norman Yoffee (Cambridge: Cambridge University Press, 2010), 45–70.

49 Sophia Perdikaris and Thomas H. McGovern, "Codfish and Kings,
Seals and Subsistence: Norse Marine Resource Use in the North
Atlantic," in *Human Impacts on Ancient Marine Ecosystems: A
Global Perspective*, ed. Rorben C. Rick and Jon M. Erlandson
(Berkley: University of California Press, 2008), 208.

See also W. Jeffrey Bolster, "Putting the Ocean in Atlantic History: Maritime Communities and Marine Ecology in the Northwest
Atlantic, 1500–1800," *American Historical Review* 113, no. 1
(2008): 19–47 and *The Mortal Sea: Fishing the Atlantic in the Age
of Sail* (Cambridge, MA: Harvard University Press, 2014), in which
he characterizes whalers and fishermen as "shock troops pushing
west" (47); Jeremy B.C. Jackson, Karen E. Alexander, and Enric
Sala, eds., *Shifting Baselines: The Past and Future of Ocean Fisheries* (Washington, DC: Island Press, 2011); Stephen Hornsby,
*British Atlantic, American Frontier: Spaces of Power in the Early
Modern British America* (Hanover: University Press of New England, 2005), 26–73.

50 UNESCO, "The Criteria for Selection," emphasis mine, accessed 2
March 2017, http://whc.unesco.org/en/criteria/. "Saga of the sea"
refers to an iconic series taken by Nova Scotia photographer W.A.

MacAskill, featuring an old sailor gesturing toward a lighthouse. See Nova Scotia Archives, http://novascotia.ca/archives/MacAskill/archives.asp?ID=3433. On the commemorative and environmental history of Lunenburg, see Claire Campbell, "Global Expectations, Local Pressures: Some Dilemmas of a World Heritage Site," *Journal of the Royal Nova Scotia Historical Society* 11, no. 1 (2008): 69–88; the chronology of commemoration is discussed in Roy Eugene Graham, FAIA and Associates, *Lunenburg, Nova Scotia: World Heritage Community Strategy* (Washington, DC: FAIA and Associates, 1998). On the state of the fishery – such as it was – in the 1990s, see Dean Bavington, *Managed Annihilation: An Unnatural History of the Newfoundland Cod Collapse* (Vancouver: University of British Columbia Press, 2010); Craig T. Palmer, "A Decade of Uncertainty and Tenacity in Northwest Newfoundland," in *Retrenchment and Regeneration in Rural Newfoundland*, ed. Reginald Byron (Toronto: University of Toronto Press, 2003), 43–64.

Interpretation at Red Bay and Battle Harbour attributes the decline of both communities to factors *unrelated* to harvesting and species decline (e.g., the loss of Basque ships commandeered by the 1588 Armada; the loss of the Battle Harbour hospital to fire in 1930, and provincial relocation programs). It is worth noting that Newfoundland was also home to the Great Auk, hunted to extinction by the 1840s, in Funk Island (now an ecological preserve).

51 Ironically, ICOMOS accepted the more critical half of the criterion – that Red Bay demonstrates "human pressure on the natural stocks of whales in the region," but not that a site of seventy-years duration constituted a "long-term example of the traditional exploitation of a marine resource." ICOMOS, "Advisory Body Evaluation for Red Bay" (UNESCO, 2013), 154–6, http://whc.unesco.org/archive/advisory_body_evaluation/1412.pdf; UNESCO, *World Heritage Nomination for the Red Bay Basque Whaling Station* (Newfoundland and Labrador: UNESCO, 2012), 92–4, http://whc.unesco.org/uploads/nominations/1412.pdf. In the management plan and on the Red Bay National Historic Site website, Parks Canada emphasizes the successive settler cultures, presumably to align with a nationalist message of Canada's history as multicultural and diverse in origin.

Parks Canada's expertise in underwater archaeology was developed in large part from working on the wrecks at Red Bay between 1978 and 1985. It then transferred these efforts to the high Arctic in search of the Franklin Expedition. See Robert Grenier, Willis Stevens, Marc-Andre Bernier, *The Underwater Archaeology of Red Bay: Basque Shipbuilding and Whaling in the 16th Century* (Ottawa: Parks Canada, 2007).

52 Sharon Babaian, "Evidence from a Disaster: The 'Ocean Ranger' Collection at the Canada Science and Technology Museum," *Material History Review* 61 (2005): 69–78.

53 "Current climate and energy policy debates in the United States rarely involve historians. If you search the Intergovernmental Panel on Climate Change's 2007 synthesis report, for instance, you will not find the words 'history' or 'historical.'" Paul Sabin, "The Ultimate Environmental Dilemma": Making a Place for Historians in the Climate Change and Energy Debates," *Environmental History* 15, no. 1 (2010): 79–80.

54 Parks Canada, *L'Anse aux Meadows National Historic Site Management Plan*, 1. For a discussion of other Viking tourism, see C. Halewood and K. Hannam, "Viking Heritage Tourism – Authenticity and Commodification," *Annals of Tourism Research* 28, no. 3 (2001): 565–80.

55 See Claire Campbell and Robert Summerby-Murray, Introduction to *Land and Sea: Environmental History in Atlantic Canada* (Fredericton: Acadiensis Press, 2013); Ian McKay, *Quest of the Folk: Antimodernism and Cultural Selection in Twentieth-Century Nova Scotia* (Montreal and Kingston: McGill-Queen's Press, 1994); C.J. Taylor, *Negotiating the Past: The Making of Canada's National Historic Parks and Sites* (Montreal and Kingston: McGill-Queen's University Press, 1990). On Louisbourg, see J.R. Donald, *The Cape Breton Coal Problem* (Ottawa: Duhamel, 1966); Meaghan Beaton, "'I Sold It as an Industry as Much as Anything Else': Nina Cohen, the Cape Breton Miners' Museum and Canada's 1967 Centennial Celebrations," *Journal of the Royal Nova Scotia Historical Society* 13 (2010): 41–62.

56 "Old Viking Village Draws 10,000 Tourists," *Montreal Gazette*, 6 September 1980. The 1975 development plan had predicted this

figure (at a time when about two thousand people visited each
year), although it was rather more optimistic and less accurate in
forecasting an average of fifty-one thousand (up to seventy-five
thousand) by the mid-1990s. H.J. Porter and Associates, "Preliminary Development Concept," 36.

57 Parks Canada, "Parks Canada Attendance 2011–12 to 2015–16:
National Historic Sites & Other Designations," last modified 31
August 2016, http://www.pc.gc.ca/eng/docs/pc/attend/table2.aspx.
See also the annual Newfoundland and Labrador Provincial
Tourism Performance reports and the 2011 Tourism Exit Survey –
Program Highlights, at http://www.btcrd.gov.nl.ca/Tourism
/tourism_research/stats/index.html. According to the national census, the Great Northern Peninsula showed a decline of 12.6% between 2001 and 2006, compared to a province-wide decline of
1.5%; and unemployment rates (36.5%) nearly *twice* that of the
province (18.5%). Ryan Gibson, "Regional Profile of the Northern
Peninsula Region, Newfoundland," Canadian Regional Development: A Critical Review of Theory, Practice, and Potentials
research project, 2013, http://cdnregdev.ruralresilience.ca/wp-
content/uploads/2013/05/NorthernPeninsulaProfile.pdf.

58 Carol Corbin is critical of Parks Canada's inability and/or unwill-
ingness to integrate into the community, and accordingly the very
limited benefits to the local population, in "Symbols of Separation:
The Town of Louisbourg and the Fortress of Louisbourg," *Environments* 24, no. 2 (1996): 15–27.

59 One study has observed that "the frequent use of blatantly inau-
thentic Viking attractions" means there are few if any references to
other cultural histories on the Great Northern Peninsula (e.g., Indigenous or Basque); the disproportionate presence of the Norse
theme succeeds in part because of the preexisting appeal in the
popular imagination but also in part because there are no "real" or
"living" populations to contest any misleading or inaccurate interpretations. Craig T. Palmer, Benjamin Wolff, and Chris Cassidy,
"Cultural Heritage Tourism along the Viking Trail: An Analysis
of Tourist Brochures for Attractions on the Northern Peninsula of
Newfoundland," *Newfoundland and Labrador Studies* 23, no. 2
(2008): 224–5.

60 When Norstead opened in 2000, the provincial government help-
    fully provided a media release titled "Differences between L'Anse
    aux Meadows National Historic Site Encampment and Norstead
    Information Sheet" (see http://www.releases.gov.nl.ca/releases/
    2000/tcr/0427bg07.htm). See also the Viking Trail Tourism Initia-
    tive Backgrounder at http://www.releases.gov.nl.ca/releases/1999/
    exec/1119n10k.htm. Norstead received further federal and provin-
    cial funding in 2007. The critique comes from James Overton in
    "A Future in the Past?: Tourism Development, Outport Archaeol-
    ogy, and the Politics of Deindustrialization in Newfoundland and
    Labrador in the 1990s," *Urban History Review/Revue d'histoire
    urbaine* 35, no. 2 (2007): 60–74. Kevin McAleese, though, suggests
    locals have embraced the Viking identity, in "L'Anse aux Meadows:
    Rediscovered and Remade."

61 Parks Canada, *L'Anse aux Meadows National Historic Site Man-
    agement Plan,* 36–7; Wallace, "An Abandoned Experiment," 219.
    The 1975 agreement that transferred the land from the province to
    Ottawa included a provision that any unused land would return
    to the province, suggesting St John's was loath to give up any more
    property than was necessary (Parks Canada, *Management Plan,*
    28–30).

62 Jerry Bannister, "A River Runs through It: Churchill Falls and the
    End of Newfoundland History," *Acadiensis* 41, no. 1 (2012): 211–
    25; Claire Campbell, "Privileges and Entanglements: Lessons from
    History for Nova Scotia's Politics of Energy," *Acadiensis* 42, no. 2
    (2013): 114–37.
        Postcolonialist academic critiques of World Heritage Sites often
    suggest that international designation and national nominations
    tend to overrule local interests and the point of view of those who
    live and work in these cultural landscapes. See, for example, the
    essays in *The Politics of World Heritage: Negotiating Tourism
    and Conservation,* ed. David Harrison and Michael Hitchcock
    (Clevedon, UK: Channel View Publications, 2005). Given the con-
    stitutional provisions for land and resource rights, as well as the
    prevailing tradition of Canadian political economy and cultural
    identity as "hewers of wood and drawers of water," I would

venture to say that another voice with a bit of distance from the incentives of resource revenues might be a good thing.

63 Artifacts found onsite include iron or wood waste from smelting, wood working, and boat repair. See Wallace, "The Norse in Newfoundland," 17–19; and "An Abandoned Experiment," 223–4.

64 See especially the essays by Alan MacEachern and John Sandlos in *A Century of Parks Canada*. To be fair, the current programs at L'Anse aux Meadows, such as boat tours, textile classes, and story-telling evenings, can be seen as reasonable extensions of the site's actual history. The Fortress of Louisbourg, in contrast, has opened up the site to all manner of events, from rock concerts to LGBTQ pride days to a gathering of hundreds of motorbikes. While this may connect the site to new community partners and attract new visitors, it's not clear if these different *uses* result in greater *understanding* of the site's history. There are also programs such as geo-caching that Parks Canada has attempted to instigate at historic sites and parks nation-wide.

65 Jerry Bannister and Roger Marsters have argued that landscape is more prominent than history in tourism/brand marketing in the Atlantic provinces in recent decades. "The Presence of the Past: Memory and Politics in Atlantic Canada since 2000," in *Shaping an Agenda for Atlantic Canada*, ed. Donald Savoie and John Reid (Halifax: Fernwood, 2011), 111–31.

66 David Roberts, "Raw and Reticent: Wilderness, Maritime Culture, Endure in Newfoundland," *Lewiston Journal*, 29 April 1989.

67 On historic parks, see Taylor, *Negotiating the Past*, specifically 28–9, 145; also Taylor, "Continuing Education: My Life as a His-torian," *Canadian Historical Review* 94, no. 1 (2013): 123; E.J. Hart, *J.B. Harkin: Father of Canada's National Parks* (Edmonton: University of Alberta Press, 2010), 140–5, especially citing Har-kin's *Report of the Commissioner of Dominion Parks* (1913), 11.

68 H.J. Porter and Associates, "Preliminary Development Concept," 4; Gimbarzevsky, *L'Anse aux Meadows National Historic Park*, 89. Having other activities such as hiking and "nature studies" was partly an acknowledgment of the "isolation and remoteness" of the

site and the effort it would take visitors to get there. In response, those that cautioned against making the site *too* much a "park" included Schonback and Dykes, "L'Anse aux Meadows National Historic Park," 2, and H.J. Porter and Associates, "Preliminary Development Concept," 29.

69 H.J. Porter and Associates, "Preliminary Development Concept," 2; Schonback and Dykes, "L'Anse aux Meadows National Historic Park," 2, 11; *L'Anse aux Meadows National Historic Park Interpretation Programme*, 6.

70 Schonback and Dykes, "L'Anse aux Meadows National Historic Park: Interpretive Plan and Exhibit Storyline," 1; H.J. Porter and Associates, "Preliminary Development Concept," 6. This is repeatedly mentioned in the 1974 and 1975 development plans.

71 Parks Canada, *L'Anse aux Meadows National Historic Site Management Plan*, 26, 32.

72 Historically marine areas have been underrepresented in conservation schemes – Canada currently has only *four* marine conservation areas. To their credit, these four all have strongly cultural/historical dimensions. The UNESCO program of marine heritage acknowledges human actors (as a cause of ecological degradation) but reserves designation for "exceptional natural phenomena" (geological features, biodiversity, etc.). John Lien and Robert Graham, eds., *Marine Parks and Conservation: Challenge and Promise*, vol. 2 (Toronto: National and Provincial Parks Association of Canada, 1985); Ameer Awad Abdulla, David Obura, Bastian Bertzky, and Yichuan Schi, *IUCN Thematic Study: Marine Natural Heritage and the World Heritage List* (Gland, Switzerland: IUCN, 2013), https://cmsdata.iucn.org/downloads/marine_natural_heritage_and_the_world_heritage_list.pdf.

CHAPTER TWO

1 UNESCO World Heritage List, "Landscape of Grand Pré," accessed 14 February 2017, http://whc.unesco.org/en/list/1404; Canadian Register of Historic Places, "Grand Pré Rural District National Historic Site of Canada," accessed 22 February 2017, http://www.historicplaces.ca/en/rep-reg/place-lieu.aspx?id=15751&pid=0. The municipality and the pré are referred to as Grand Pré, while

the specific national historic site is the Grand-Pré National Historic Site.

2 W. Eilers, R. MacKay, L. Graham, and A. Lefebvre, eds., *Environmental Sustainability of Canadian Agriculture: Agri-Environmental Indicator Report Series – Report no. 3* (Ottawa: Agriculture and Agri-Food Canada, 2010). The County of Kings, which encompasses the eastern half of the Annapolis Valley, has the most agriculturally based economy in Nova Scotia, in 2001 accounting for 30 percent of the province's agriculture and 2.5 times the national average per capita. Nomination Grand Pré, *Management Plan for the Landscape of Grand Pré* (January 2011), 28, http://nomination grandpre.ca/Dossier/2A%20-%20management%20plan%20land scape%20of%20GP.pdf.

3 On the former, see especially the work of Naomi Griffiths, such as *From Migrant to Acadian: A North American border people, 1604–1755* (Montreal and Kingston: McGill-Queen's University Press, 2005); and John G. Reid, *The "Conquest" of Acadia, 1710: Imperial, Colonial, and Aboriginal Constructions* (Toronto: University of Toronto Press, 2004); and John G. Reid with Emerson W. Baker, *Essays on Northeastern North America, Seventeenth and Eighteenth Centuries* (Toronto: University of Toronto Press, 2008). On the latter, see Ian McKay and Robin Bates, *The Province of History: The Making of the Public Past in Twentieth-Century* (Montreal and Kingston: McGill-Queen's University Press, 2010); also Monica MacDonald, "Railway Tourism in the 'Land of Evangeline,' 1882–1946," *Acadiensis* 35, no. 1 (2005): 158–80; M. Brook Taylor, "The Poetry and Prose of History: Evangeline and the Historians of Nova Scotia," *Journal of Canadian Studies* 23, no. 1–2 (1988): 46–67.

4 Examples include Jonathan Fowler, "A Walk back in Time at Grand Pré," in *Underground Nova Scotia: Stories of Archaeology,* eds. Paul Erickson and Jonathan Fowler (Halifax: Nimbus Publishing, 2010), 43–54; J. Sherman Bleakney, *Sods, Soil, and Spade: The Acadians at Grand Pré and Their Dykeland Legacy* (Montreal and Kingston: McGill-Queen's University Press, 2004); Graeme Wynn, "Late Eighteenth-Century Agriculture on the Bay of Fundy Marshlands," *Acadiensis* 8, no. 2 (1979): 80–9; Matthew Hatvany,

"'Wedded to the Marshes: Salt Marshes and Socio-Economic Dif-
ferentiation in Early Prince Edward Island," *Acadiensis* 30, no. 2
(2001): 40–55 and "The Origins of the Acadian 'Aboiteau': An
Environmental-Historical Geography of the Northeast," *Historical
Geography* 30 (2002): 121–37; Karl W. Butzer, "French Wetland
Agriculture in Atlantic Canada and Its European Roots: Different
Avenues to Historical Diffusion," *Annals of the Association
of American Geographers* 92, no. 3 (2002): 451–70; Gregory
Kennedy, "Marshland Colonization in Acadia and Poitou during
the 17th Century," *Acadiensis* 42, no. 1 (2013): 37–66 and *Some-
thing of a Peasant Paradise?: Comparing Rural Societies in Acadie
and the Loudunais, 1604–1755* (Montreal and Kingston: McGill-
Queen's University Press, 2014).

5 Here I think of the debate about the relative prosperity in mid-
nineteenth century Nova Scotia between scholars such as Julian
Gwyn, most fully in *Excessive Expectations: Maritime Commerce
and the Economic Development of Nova Scotia* (Montreal and
Kingston: McGill-Queen's University Press, 1998), and others such
as Kris Inwood and Phyllis Wagg in "Wealth and Prosperity in
Nova Scotian Agriculture, 1851–71," *Canadian Historical Review*
75, no. 2 (1994): 239–64. Margaret Conrad's article on the apple
industry is also in this vein, characteristic of the "Acadiensis
School" of Maritime historiography: "Apple Blossom Time in the
Annapolis Valley, 1880–1957," *Acadiensis* 9, no. 2 (1980): 14–39.
The origins of the Acadiensis School – which brought regional history to the fore, but with a strong interest in the political economy
of Confederation – is described by Del Muise in "Organizing Historical Memory in the Maritimes: A Reconnaissance," *Acadiensis*
30, no. 1 (2000): 50–60. For more on the political economy of this
era, see Daniel Samson, *The Spirit of Industry and Improvement:
Liberal Government and Rural-Industrial Society, Nova Scotia,
1790–1862* (Montreal and Kingston: McGill-Queen's University
Press, 2008).

6 I say this with two rather substantial qualifiers: the opening line,
"In the forest primeval" is not an accurate description of the Fundy
marshlands, and, of course, the entire storyline of the poem is

fictitious. Henry Wadsworth Longfellow, *Evangeline: A Tale of Acadie* (Boston: William D. Ticknor & Company, 1847).

7 By way of illustration, see Julian Gwyn, "Shaped by the Sea but Impoverished by the Soil: Chester Township to 1830," *The Nova Scotia Planters in the Atlantic World, 1759–1830*, ed. T. Stephen Henderson and Wendy G. Robicheau (Fredericton: Acadiensis Press, 2012), 99–121.

8 In the *Grænlendinga Saga*, Vinland provides not only fruit and wheat but water ("the sweetest thing they had ever tasted") and "bigger salmon than they had ever seen." Not surprisingly, "the country seemed to them so kind." In Magnus Magnusson and Hermann Pálsson, eds., *The Vinland Sagas: The Norse Discovery of America* (London: Penguin Classics, 1965), 55–6. On the theory of Vinland in Nova Scotia, see Mats G. Larsson, "The Vinland Sagas and the Actual Characteristics of Eastern Canada: Some Comparisons with Special Attention to the Accounts of the Later Explorers," in *Vinland Revisited: The Norse World at the Turn of the First Millennium,* ed. Shannon Lewis-Simpson (St John's: Historic Sites Association of Newfoundland and Labrador, 2003), 391–8.

9 Paul Mascarene, "Description of Nova Scotia," written for the Board of Trade, 1720, in Thomas B. Akins, *Selections from the Public Documents of the Province of Nova Scotia* (C. Annand, 1869), 46.

10 Alan R. MacNeil, "Early American Communities on the Fundy: A Case Study of Annapolis and Amherst Townships, 1767–1827," *Agricultural History* 63, no. 2 (1989): 101–19; Debra McNabb, "The Role of Land in the Development Horton Township, 1760–1775," in *They Planted Well: New England Planters in Maritime Canada*, ed. Margaret Conrad (Fredericton: Acadiensis Press, 1988), 151–60. At the head of the Bay of Fundy, on what is now the border of Nova Scotia and New Brunswick, the British also invited settlers from Yorkshire with experience in drainage farming. Nelson Bezanson and Robert Summerby-Murray, *Background Report: An Historical Geography of Tantramar's Heritage Landscapes* (Fredericton: Department of Municipalities, Culture and Housing, Heritage Branch, 1996).

11 "Resolution of the Chamber of Commerce of Halifax," 9 July
    1825, reprinted in Charles R. Fairbanks, *Reports and papers relat-
    ing to a canal intended to connect the harbour of Halifax with the
    Basin of Mines: Remarks on its nature and importance, and a plan
    and section: Also, the report of a survey for canals between St.
    Peter's Bay and the Bras d'Or Lake in Cape Breton, and the Bay of
    Fundy and Bay of Verte* (Halifax, 1826), 30. The same source cites
    correspondence from Lieutenant-Governor John Wentworth as
    early as 1798, urging the building of a canal "To [enhance] the
    commerce of the Province, by facilitating – rendering safe – and
    reducing the expense of all commodities and produce, upon that
    extensive fertile and productive country to which such a Canal,
    as is proposed, immediately communicates" (12).

12 William D. Lawrence, nomination day speech, 1863, in *William D.
    Lawrence: Nova Scotia Shipbuilder & Anti-Confederation Cam-
    paigner, The Complete Archived and Annotated Writings* (Kennet-
    cook, NS: Heroes of Hants County Association, 2010), 205.

13 Graeme Wynn, "'Deplorably Dark and Demoralized Lumberers'?:
    Rhetoric and Reality in Early Nineteenth-Century New Bruns-
    wick," *Journal of Forest History* 24, no. 4 (1980): 168–76.
    Graeme also shaped my thinking with his discussion of scale in
    "Thinking about Mountains, Valleys and Solitudes: Historical
    Geography and the New Atlantic History," *Acadiensis* 30, no. 1
    (2001): 129–45. On the heroic Planter historiography, see R.S.
    Longley, "The Coming of the New England Planters to the An-
    napolis Valley," *Collections of the Royal Nova Scotia Historical
    Society* (1960), 33, reprinted in Conrad, *They Planted Well*; also
    Alison Norman, "'A Highly Favoured People': The Planter Narra-
    tive and the 1928 Grand Historic Pageant of Kentville, Nova
    Scotia," *Acadiensis* 38, no. 2 (2009): 116–40.

14 Prescott House, "Mandate and Mission," accessed 22 February
    2017, https://prescotthouse.novascotia.ca/about-prescott-house/
    mandate-and-mission; Canadian Register of Historic Places,
    "Prescott House," accessed 14 February 2017, http://www.historic
    places.ca/en/rep-reg/place-lieu.aspx?id=1499&pid=0. On the apple
    industry see Keith Hatchard's six-part "The History of the Apple

Industry of Nova Scotia," *Nova Scotia Historical Quarterly*
(1977–78) 7, no. 1–4; 8, no. 1 and 3.

15 R.G. Haliburton, "The Past and Future of Nova Scotia: An Address on the 113th Anniversary of the Settlement of the Capital of
the Province," in *Nova Scotia in 1862: Papers Relating to the Two
Great Exhibitions in London in That Year* (Halifax: T. Chamber-
lain, 1864), 24.

16 Joseph Howe, "Mr. Howe on Confederation," 1866, in *The
Speeches and Public Letters of Joseph Howe (Based upon Mr. An-
nand's Edition of 1858)*, ed. Joseph Andrew Chisholm (Halifax:
Chronicle Publishing Co. Ltd., 1909), 476. On the "global" stand-
ing and mentality of Nova Scotia in this era, see Tom Peace, Jim
Clifford, and Judy Burns, "Maitland's Moment: Turning Nova
Scotia's Forests into Ships for the Global Commodity Trade in the
Mid-Nineteenth Century," in *Moving Natures: Mobility and Envi-
ronment in Canadian History*, ed. Ben Bradley, Jay Young, and
Colin Coates (Calgary: University of Calgary Press, 2016), 27–54.

17 For the fuller story of this see Conrad, "Apple Blossom Time."
While the earliest railways in Nova Scotia were geared toward coal
deposits, by the mid-nineteenth century the Annapolis Valley was
crisscrossed by several lines including the Nova Scotia Railway,
Windsor–Annapolis Railway, Western Counties Railway, Cornwallis Valley Railway, Nova Scotia Central Railway, and the Middleton
and Victoria Beach Railway. Most important were the Intercolonial
Railway (a condition of Confederation and completed to Quebec
in 1879) and the Dominion Atlantic Railway (the result of a
merger between the Windsor–Annapolis Railway and Western
Counties Railway). See David E. Stephens, *Iron Road: Railways
of Nova Scotia* (Halifax: Lancelot Press, 1972).

18 Alisa Smith and J.B. Mackinnon, *The 100 Mile Diet: A Year of
Local Eating* (Toronto: Random House, 2007), 96–7.

19 "For Better Fruit," *Berwick Register*, 20 January 1910.

20 Now the Atlantic Food and Horticulture Research Centre, an
overview of its history can be found in Atlantic Food and Horticulture Research Centre, "Kentville's Century of Science," cat. no.
A52-186/2011E, Agriculture and Agri-Food Canada, 2011,

http://publications.gc.ca/collections/collection_2011/agr/A52-186-2011-eng.pdf. Its publications, often with scholars from the Nova Scotia Agricultural College, are prolific.

21 The Windsor–Annapolis Railway opened its line to Grand Pré in 1869. See especially Barbara LeBlanc, *Postcards from Acadie: Grand-Pré, Evangeline and the Acadian Identity* (Kentville, NS: Gaspereau Press, 2003); Jay White, "'A Vista of Infinite Development': Surveying Nova Scotia's Early Tourism Industry," *Collections of the Royal Nova Scotia Historical Society* 6 (2003): 144–69.

22 Charles H. Towne, *Ambling through Acadia* (New York and London: The Century Co., 1923), 145; Dominion Atlantic Railway, "The Land of Evangeline, Nova Scotia," pamphlet, 1931; "Nova Scotia by Canada Pacific," pamphlet, n.d., 11; Yarmouth Steamship Company, "Beautiful Nova Scotia," pamphlet, 1894, 33; Longfellow, *Evangeline: A Tale of Acadie*. See *"Canada's Ocean Playground": The Tourism Industry in Nova Scotia, 1870–1970*, virtual exhibit by the Nova Scotia Archives, https://novascotia.ca/archives/tourism/.

23 Ian McKay, "Cashing in on Antiquity: Tourism and the Uses of History in Nova Scotia, 1860–1960," in *Settling and Unsettling Memories: Essays in Canadian Public History*, ed. Nicole Neatby and Peter Hodgins (Toronto: University of Toronto Press 2012), 454–90. McKay's characterization of interwar Nova Scotians as "unwillingly post-industrial" (464) is telling; it acknowledges the end or exhaustion of staple resources, and the resulting economic fragility, but also the desire to *continue* in the industrial vein.

Marshes and wetlands were generally seen as an impediment to modernity or a symbol of a pre-modern era, and as such were mapped, gridded, and drained to facilitate built development. Here, in contrast, marshland could be seen as pastoral and productive. Rodney J. Giblett, *Postmodern Wetlands: Culture, History, Ecology* (Edinburgh: Edinburgh University Press, 1996), 12.

24 "Watching the Apples Grow," Stan Rogers (SOCAN) c p 1976, *Fogarty's Cove*. Used by permission. The song refers to the town of Wolfville (to the west of Grand Pré) and Gaspereau Mountain (to the east). (My thanks to the King's College chapel choir members

who taught me this song on the ferry to Prince Edward Island as a
first-year undergraduate.) On the highway geography of the interwar period, see Sarah Osborne, "The Road to Yesterday: Nova
Scotia's Tourism Landscape and the Automobile Age, 1920–1940"
(master's thesis, Dalhousie University, 2009); and Alan MacEachern, *Natural Selections: National Parks in Atlantic Canada* (Montreal and Kingston: McGill-Queen's Press, 2001).

25 Windsor is home to the Hants County Exhibition, the oldest agricultural fair in Canada; and was also home to Howard Dill, who
gained international recognition for the Valley's pumpkins with
his breed of "Atlantic Giant."

26 This is how Grand Pré Wines describes the white varietal "Tidal
Bay" on their website; also see the language about their British reception in Brett Bundale, "Benjamin Bridge Bubbly to be Uncorked
during Olympics," *Chronicle-Herald* (Halifax), 24 July 2012.
There are now nine wineries in the area, and two more at the far
end of the Annapolis Valley at Bear River.

27 This is how Nova Scotia Crystal describes its "Annapolis" pattern
on its website.

28 Craig Flinn quoted in "Top Chefs," *The Coast*, 13 September
2012.

29 Statement of Commemorative Intent for Grand Pré National His-
toric Site drafted based on the resolution by the Historic Sites and
Monuments Board (HSMBC) in 1982, in Parks Canada, *Grand-Pré
National Historic Site of Canada Management Plan* (Ottawa: Her
Majesty the Queen in Right of Canada, 2002), especially Appendix
1 for the evolving designations by the HSMBC from 1955 through
to 1982. On cultural communities, see Parks Canada, *National
Historic Sites of Canada System Plan* (Ottawa: Her Majesty the
Queen in Right of Canada, 1997). In the 1920s, Acadian community responded to the DAR's Evangeline with a memorial cross and
a replica of St Charles Church, styles of memorial grounded in
their historical values rather than literary fiction. Ottawa acquired
the site in 1957, but it was only in the early 1980s (the HSMBC
designation in 1982, and a new management plan in 1985) that
the Acadian cultural claim became central. "Ancestral homesite" is
now used to describe both Grand Pré and the Melanson Settlement

on the Annapolis River. See Parks Canada's online Directory of
Federal Heritage Designations at http://www.pc.gc.ca/apps/dfhd/
default_eng.aspx. On the site's commemorative history and its
relationship with the Acadian community, see Michael Gagné,
"'Memorial Constructions': Representations of Identity in the
Design of the Grand-Pré National Historic Site, 1907–Present,"
Acadiensis 42, no. 1 (2013): 67–98. The Melanson Settlement in
Annapolis, Nova Scotia is a second national historic site (desig-
nated in 1987) that commemorates Acadian marshland agriculture.

30 Canadian Register of Historic Places, "Grand-Pré Rural Historic
District," 22 February 2017, http://www.historicplaces.ca/en/rep-
reg/place-lieu.aspx?id=15751, emphasis mine. See also "How Private Property Owners Can Preserve a Heritage District: The Case
of Grand-Pre Rural Historic District, Nova Scotia," in *The National Standards and Guidelines for the Conservation of Historic
Places in Canada*, 2nd ed. (Ottawa: Her Majesty the Queen in
Right of Canada, 2010), 10–12.

31 I am grateful to Conevery Valencius and the other members of the
Northeast and Atlantic Canada Environmental History Forum
from 2012 for their thoughts on this point.

32 UNESCO, "History and Terminology, Cultural Landscapes," accessed 24 February 2017, http://whc.unesco.org/en/culturalland
scape/.

33 Graeme Wynn, "Reflections on the Environmental History of
Atlantic Canada," in *Land and Sea: Environmental History in
Atlantic Canada*, ed. Claire Campbell and Robert Summerby-
Murray (Fredericton: Acadiensis Press, 2013), 239.

34 A.J.B. Johnston, "Imagining Paradise: The Visual Depiction of
Pre-Deportation Acadia, 1850–2000," *Journal of Canadian Studies*
38, no. 2 (2004): 122–3.

35 Conrad, "Apple Blossom Time," 39.

36 Nomination Grand Pré, *Management Plan for the Landscape of
Grand Pré*, 27–35. Also Bill Freedman, Michael Macdonald, and
Harry Beach, *Ecological Conditions at the Grand-Pré National
Historic Site* (Halifax: Parks Canada, 2001) and *Grand-Pré Na-
tional Historic Site of Canada State of the Site Report* (Kouchi-

bouguac, NB: Parks Canada, 2009). I thank my colleague, Nicholas
Hill, for teaching me about the salt marsh ecology at Grand Pré
and the Bay of Fundy.

37 Green Gables is part of L.M. Montgomery's Cavendish National
Historic Site. As Parks Canada writes on the Cavendish website,
the Green Gables house has been "restored … to reflect a typical
farm house of the late Victorian period [and] with the reconstruction of period farm outbuildings, namely a barn, granary and
woodshed." My thoughts on Prince Edward Island have been
shaped by Matthew Hatvany, "Maps, History, and Environmental
Histories in PEI Marshfield," presented at Time and a Place: Environmental Histories, Environmental Futures, and Prince Edward
Island, University of Prince Edward Island, 17 June 2010, http://
niche-canada.org/resources/conference-workshop-archive/time-and-
a-place-environmental-histories-environmental-futures-and-prince-
edward-island/. Also Colin MacIntyre, "The Environmental Pre-
History of Prince Edward Island 1769–1970: A Reconnaissance in
Force" (master's thesis, University of Prince Edward Island, 2010).

38 Here I am thinking of Dean Bavington, *Managed Annihilation:
An Unnatural History of Newfoundland Cod Collapse* (Vancouver:
University of British Columbia Press, 2010); and Jeremy B.C. Jack-
son, Karen E. Alexander, and Enric Sala, eds., *Shifting Baselines:
The Past and the Future of Ocean Fisheries* (Washington: Island
Press, 2011). See also Claire Campbell, "Global Expectations,
Local Pressures: Some Dilemmas of a World Heritage Site," *Jour-
nal of the Royal Nova Scotia Historical Society* 11, no. 1 (2008):
1–18. "Saga of the sea" refers to a popular series of photographs
by W.A. MacAskill (ca. 1936) of a fisherman and boy at Peggy's
Cove lighthouse; see Nova Scotia Archives, http://novascotia.ca
/archives/MacAskill/archives.asp?ID=3433.

CHAPTER THREE

1 G. Brian Woolsey, Historic Sites Branch, to Dr. Fred Armstrong,
University of Western Ontario, 26 September 1972, file F-NA-65,
RG 47-64 77-455, box 1, Archives of Ontario (hereafter AO). Jean
F. Morrison, the staff historian at Fort William for fifteen years,

wrote an accessible history of the fort in *Superior Rendez-Vous Place: Fort William in the Canadian fur trade*, 2nd ed. (Toronto: Natural Heritage Books, 2007).

2 Washington Irving, *Astoria: Or, Enterprise beyond the Rocky Mountains* (London: Richard Bentley, 1839) 7; Robert Englebert, "Merchant Representatives and the French River World, 1763–1803," *Michigan Historical Review* 34, no. 1 (2008): 63–82.

3 David Kemp, "The Impact of Weather and Climate on the Fur Trade in the Canadian Northwest," originally published Thunder Bay Historical Museum Papers & Records, vol. 8 (1980), reprinted in *Lake Superior to Rainy Lake: Three Centuries of Fur Trade History, A Collection of Writings*, ed. Jean Morrison (Thunder Bay: Thunder Bay Historical Museum Society, 2003), 32–42; also Eric C. Morse, *Fur Trade Canoe Routes of Canada, Then and Now* (Ottawa: Queen's Printer, 1969), 27–32. While it took six to eight weeks to travel upriver from Montreal to Fort William, the return trip was much faster: three weeks to a month. Carolyn Podruchny, *Making the Voyageur World: Travelers and Traders in the North American Fur Trade* (Lincoln: University of Nebraska Press, 2006), 101.

4 Gabriel Franchère, *Narrative of a Voyage to the Northwest Coast of America in the Years 1811, 1812, 1813, and 1814 or the First American Settlement on the Pacific*, ed. and trans. J.V. Huntington (New York: Redfield, 1854), 339.

5 D'Arcy Jenish, *Epic Wanderer: David Thompson and the Mapping of the Canadian West* (Lincoln: University of Nebraska Press, 2003), 8, 211.

6 The city of Thunder Bay was created in 1970 through the administrative merger of two cities on either side of the McIntyre River: Port Arthur to the north and Fort William to the south.

7 The 1923 designation emphasized the role of the site in a series of transcontinental *events*, including the construction of two transcontinental railways and the first arrival of grain from the prairies. HSMBC plaque, Thunder Bay, Ontario. The fort as a *site* of national historic importance was designated in 1968.

8 "Key Link in Days of East–West Fur Trading, Old Fort William to Rise Again at Lakehead," *Globe and Mail*, 21 January 1971, 4.

In 1971, National Heritage Limited appeared a fairly respectable choice. Incorporated in 1969, it was no fly-by-night operation – clients included the Ontario Northland Transportation Commission, the Manitoba Historical Society, and the United States National Parks Service. For *Hinge of a Nation*, it consulted such prominent historians as Maurice Careless, Morris Zaslow, Fred Armstrong, Alan Gowans, J.I. Rempel, and Ronald Way, a leading authority on Old Fort Henry and Upper Canada Village. After 1971, National Heritage Limited would also hire a number of younger researchers who represented a new voice from the emerging field of public history. Ah, those were the days, when younger researchers could *get* hired. See National Heritage Limited, *Life, Steam and Iron* (North Bay: Ontario Northland Transportation Commission, 1974); Edith Steenson, Honorary Secretary's Report, *Manitoba Historical Society Transactions* 3, no. 27 (1970–71); John A. Hussey, *Fort Vancouver: Historic Structures Report: Historical Data*, vol. 2 (Denver: National Park Service, 1976), chapter 1, n96. The researchers included Peter Pratt, Norman Ball, Ron Stagg, Ron Way, Joan Halloran, and Michelle Greenwald. Greenwald, however, described the older, renowned historians as "window-dressing" who lent their names but were rarely if ever consulted. Michelle Greenwald, personal correspondence (email), 19 August 2007.

In 1974, the *Globe and Mail* would run a damaging exposé on the company, questioning why it had received the $10 million contract for Fort William without tender or competition. The newspaper revealed that National Heritage had a distinct habit of borrowing officers from the Progressive Conservative party (its secretary-treasurer and director were both Members of Parliament, its vice-president of development a provincial advisor), and many of the researchers on its "rather remarkable team" had been fired or had quit (Gerald McCauliffe, "Another Ontario Contract without Tenders – $10 Million to Build Fort," 4 February 1974, 1; "A Disneyland of Northern Ontario?" *Globe and Mail*, 4 February 1974, 4.) Ronald Stagg writes, "I was one of the longest lasting employees of National Heritage. Since I was with the company about two years, you can read a lot into that" (personal correspondence [email], 3 May 2007). By July 1976, National Heritage – by this point reduced to renting a room at York University's Glendon College – had declared bankruptcy. Apparently it lost money in expanding its commercial projects, such as selling reproductions of North West Company blankets in the United States, "in an attempt to capitalize on its investment in historical research" (Arthur Johnston, "Company that Built Thunder Bay Park Owes $396,000, Folds," *Globe and Mail,* 1 July 1976, 1.) See Alan Gordon, *Time Travel: Tourism and the Rise of the Living History Museum in Mid-Twentieth-Century Canada* (Vancouver: University of British Columbia Press, 2016), 224–7. My thanks to Mark Leeming for assistance with these articles.

9 Peter H. Bennett, Assistant Director, National and Historic Parks Branch, Department of Indian and Northern Affairs, "The Recent Discovery of Canada's History," paper given 14 February 1973 in Toronto, file PA-46 Parks Canada – Historical Parks and Sites, RG 47-64, box 14, AO.

10 On Alberta, see Mark Rasmussen, "The Heritage Boom: The Evolution of Historical Resource Conservation in Alberta," *Prairie Forum* 15, no. 2 (1990): 259; Gerald George, *Visiting History: Arguments over Museums and Historic Sites* (Washington: American Association of Museums, 1990), 26. On the relationship between Fort William and Louisbourg, see Don MacLeod, Historic Sites Branch Proposal, Archaeological Publication Programme: Fort William, 1974–75, 29 April 1974, file F-AR 35, RG 47-64 77-455, box 2, AO. Peter Boyle, Manager of Historical Collections, has said that the Fort William project is "on the scale of Louisbourg" in the depth of its archival collections (phone conversation, 14 February 2007). In 1972, Woolsey and Gary Sealey visited Louisbourg along with archaeologist Donald MacLeod. They admired its historical and archaeological research program, but – interesting in this context – felt its use of surrounding landscape underdeveloped. Woolsey to Sealey, 17 August 1972, file PA-46 Parks Canada – Historical Parks and Sites, RG 47-64, box 14, AO.

Ronald Stagg and others had complained to the director of the Branch that researchers had been unable to consult other site historians or collections: "people designing and building the various structures at the Fort have not been allowed, despite repeated requests, to see either existing Red River frame buildings or any other kind of historical building in Canada, whether original, restoration or reconstruction, except for Fort Edmonton and St. Marie Among the Hurons." Stagg to John Sloan, 9 November 1972, file F-NA-65, RG 47-64 77-455, box 1, AO; also correspondence from Michelle Greenwald, 19 August 2007: "We were not allowed up to see the site ... Researchers were not allowed to visit the site, discuss findings or theories with archaeologists or other historians, etc." Since the assistant director of the National Historic Parks Branch had told Keenan that he was happy to have Ontario staff "pick our brains" about period furnishings and their work at Lower Fort Garry, a more prestigious project with extensive physical remains, passing up such an opportunity seems odd, and reinforces the suggestion that National Heritage wanted the construction done as quickly as possible. Peter H. Bennett to J.W. Keenan, 26 October 1971, file PA-46 Parks Canada – Historical Parks and Sites, RG 47-64, box 14, AO. National Heritage had stated – inaccurately – "there are no precedents in the history of construction for a project such as the reconstruction of our Fort." *Fort William: Hinge of a Nation* (Toronto: National Heritage Limited, 1970), 97.

On Louisbourg, see C.J. Taylor, *Negotiating the Past: The Making of Canada's National Historic Parks and Sites* (Montreal and Kingston: McGill-Queen's University Press, 1990), especially 169–90; Terrence MacLean, *Louisbourg Heritage: From Ruins to Reconstruction* (Sydney, NS: University College of Cape Breton Press, 1995) and "The Making of Public History: A Comparative Study of Skansen Open Air Museum, Sweden; Colonial Williamsburg, Virginia; and the Fortress of Louisbourg National Historic Site, Nova Scotia," *Material History Review* 47 (Spring 1998) 21–32; A.J.B. Johnston, *Louisbourg: Past, Present, and Future* (Halifax: Nimbus Press, 2014). Bruce Fry defends the archaeological/research program in producing a meritorious reconstruction, and feels that it is the constraints of a public site and modern environment that detract from its sensibility; as he says, "The form is there, but not the substance ... the air is no longer redolent with the odor of drying cod" (212). "Designing the Past at Fortress Louisbourg," in *The Reconstructed Past: Reconstructions in the Public Interpretation of Archaeology and History*, ed. John H. Jameson Jr. (Walnut Creek, CA: AltaMira Press, 2003), 199–214.

11 Alan Gordon, "Heritage and Authenticity: The Case of Ontario's Sainte-Marie-among-the-Hurons," *Canadian Historical Review* 85, no. 3 (2004): 520–1. On Upper Canada Village, see William T. Anderson, "Upper Canada Village: St. Lawrence River Town," *History News* 20, no. 9 (1965); for a more critical view of one town affected by the circumstances surrounding the Village's creation, see Joy Parr's profile of Iroquois, Ontario, in *Sensing Changes: Technologies, Environments, and the Everyday, 1953– 2000* (Vancouver: University of British Columbia Press, 2009).

   The need for fundraising, although hardly new, openly affected museums' public functions as well as curatorial decisions. Patricia Wood, for example, has argued that dependence on private donations at Calgary's Heritage Park altered its collections mandate to the extent that the park has sacrificed period integrity for popular appeal and good relations with potential donors. Patricia K. Wood, "The Historic Site as Cultural Text: A Geography of Heritage in Calgary, Alberta," *Material History Review* 52 (Fall 2000): 33–43; see also Robyn Gillam, *Hall of Mirrors: Museums and the Canadian Public* (Banff, AB: Banff Centre Press, 2001), especially chapter 4.

12 D.F. McOuat to Fernand Guidon, Minister of Public Records and Archives, 16 March 1971, file 3:5 Fort Albany, RG 47-64 77-455, box 7, AO; G.P. Elliott, District Manager to Woolsey, 25 June 1973, file 3:3 Chapleau, RG 47-64 77-455, box 18, AO. See also the "Historic Fur Trade Posts Survey, 1970," file 3.5 Hudson Bay Lowlands, RG 47-64 77-455, box 7, AO.

   To the west, Lower Fort Garry and Fort Langley were comprehensive and lavish reconstructions with detailed interior furnishings and costumed interpreters, chosen in part to distribute federal presence carefully across the country. But they succeeded in part because of, again, federal presence, and because of their proximity to Winnipeg and Vancouver. M.B. Payne and C.J. Taylor, "Western

Canadian Fur Trade Sites and the Iconography of Public Memory," *Manitoba History* 46 (Fall/Winter 2003–04): 6.

13 Podruchny, *Making the Voyageur World*, 165. The American National Park Service referred to Fort William as "a veritable city in the wilderness." Ted Catton, Marcia Montgomery, and Historic Research Associates, Inc., *Special History: The Environment and the Fur Trade Experience in Voyageurs National Park, 1730–1870* (Missoula, MT: National Park Service, 2000). The same image of merriment and resolve in a frontier pervades Port Royal, Samuel de Champlain's first attempt at colonization in Acadie in 1604. Here Champlain founded an "Order of Good Cheer" to boost morale during a wretched first winter, and true enough, in the *habitation* reconstructed in the late 1930s sits a lavish dining table laid for Good Cheer.

14 National Heritage, *Hinge of a Nation*, ii; Stagg, personal correspondence, 20 August 2007.

15 Gordon, *Time Travel*, 221.

16 National Heritage, *Hinge of a Nation*, 25, 62–3.

17 Brian Woolsey, Historic Sites Planner, "An Interpretative Framework for Old Fort William Historical Park," September 1972, file F-IN-77 Interpretative Program 1972, RG 47-64 77-455, box 1, p. 14, AO. Designating Mattawa House as a provincial historic site, the Department of Travel and Publicity announced that "This canoe route might well be termed Trans-Canada Highway #1," press release, 28 June 1961, file 4.2 Mattawa Wild River Park, RG 47-64 77-455, box 7, AO. On Lake Superior, Graham MacDonald to Woolsey, 12 September 1973, file 4.7 Lake Superior Provincial Park, RG 47-64 77-455, box 18, AO.

Stephen Daniels makes a similar observation about the Hudson River in American mythology, which became "aligned to an epic, imperial trajectory of national development." *Fields of Vision: Landscape Imagery and National Identity in England and the United States* (Oxford: Polity Press, 1993), 147.

18 National Heritage, *Hinge of a Nation*, Foreword, n.p.

19 Martin O'Malley, "Brief Thunder Bay Visit Provides Much Fun for Royal Couple," *Globe and Mail*, 4 July 1973, 1–2.

20 For further discussion of the critiques and controversies in this

NOTES TO PAGES 79–81

period, see Claire Campbell, "'Hinge of a Nation' or Bone of Contention: The Battle over Reconstructing Fort William" (paper presented at the Canadian Historical Association Annual Meeting,
Saskatoon, May 28–30, 2007). Jean Halloran also discusses the
status of the reconstruction in "Wooden Forts of the Early Northwest: Fort William," *Association pour la préservation et ses tech-
niques/Association for Preservation Technology Bulletin* 6, no. 2
(1974): 39–81.

21 In 1965, Ontario's Archaeological and Historic Sites Board
awarded Kenneth Dawson funds for a survey of sites on the Dawson Trail and Fort Kaministiquia at Fort William. AHSB Minutes,
23 June 1965, file 4.3 Archaeological and Historic Sites 1965, RG
5-4, box 6, AO. See also Kenneth C.A. Dawson, "A Preliminary
Archaeological Investigation of the North West Company Post on
the Kaministiquia River, Fort William, 1800–1821," manuscript
report, Minister of Tourism and Information, Old Fort William
restoration, RG 5-14, barcode B355119, AO.

22 Though the federal government, at least, had set a precedent by
appropriating private property for site development, as at Louisbourg. At Thunder Bay, estimates of the cost of acquiring the site
varied; the Minister of Tourism and Information put the price at
upwards of $784,000 (James A.C. Auld to R.B. Mann, President
of Thunder Bay, 8 February 1971, file F-AG-55, RG 47-64 77-455,
box 1, AO); National Heritage estimated expropriation costs at
$2.86 million, not to mention the price of "human suffering and
deprivation" (!) (*Hinge of a Nation,* Appendix G., n.p.). Not surprisingly, costs escalated as reconstruction progressed. By 1973, the
total price tag was already predicted to reach nearly $12.1 million
("Fort William Historical Park, Policy Submissions, 1973," file
F-PO-55, RG 47-64 77-455, box 1, AO). The working plan for revi-
talizing the downtown core of Thunder Bay made no reference to
heritage of any kind in 1969, and anticipated the railway tracks
being left in place for the foreseeable future. Proctor, Redfern, Bousfield, and Bacon, *Downtown Urban Renewal Scheme, Fort William:
Final Report* (Toronto: Redfern, Bousfield, and Bacon, 1969).

23 Rasmussen, "The Heritage Boom"; Bodo Werner and P. Hawker,

Fort Edmonton: The Reconstruction Story (Edmonton: Parks
and Recreation, 1974); "Key Link," *Globe and Mail*, 21 January
1971. Pointe de Meuron had its own history connected with Fort
William's heyday. It was here that Swiss infantry – the De Meurons
– camped in 1816, when Thomas Douglas, the Earl of Selkirk,
hired them to capture the Nor'westers' fort, in retaliation for NWC
efforts to undermine Selkirk's settlement at Red River. Afterwards
the Hudson's Bay Company operated a small and generally unsuccessful post here; it was no competition for Fort William, and effectively served a placeholder until the merger of the two companies
in 1821. See Susan J. Campbell, "Competitive Fur Trade Tactics:
Pointe de Meuron, 1817–1821," Thunder Bay Historical Museum
Society Papers and Records, vol. 1 (1973): 33–40.

24 Woolsey, "An Interpretative Framework," 4; Greenwald, personal
correspondence (email), 19 August 2007.

25 Joyce Kleinfelder, Lakehead University, to Robert Bowes, Director,
Historic Sites Branch, 21 June 1974, file F-AR 35, RG 47-64 77-
455, box 2, AO. In his *Preliminary Archaeological Investigation*,
Dawson described the challenges of excavating in a major working
rail yard: working around trucks and active rail lines, or finding
their stakes removed by CPR workers. From the start, National
Heritage seemed to view the archaeological project useful only in-
sofar as it confirmed their preexisting site plan. *Hinge of a Nation*
– and National Heritage Limited president, William Piggott –
stated that their approach was to accept Selkirk unless proved
wrong by archaeological evidence. But the archaeologists disliked
hunting for features on the Selkirk plan, and would have preferred
a more systematic excavation. National Heritage, *Hinge of a Nation*, 49; Minutes of Meeting [of Fort William Project Committee],
14 September 1972, file Fort William Archaeology 1972; "Reorien-
tation of the Fort William Archaeological Project," 1973?, file Fort
William Archaeology 1973; "Fort William Archaeological Project
1974–75 Program," from Kleinfelder, 5 February 1974, file F-AR
35. All RG 47-64 77-455, box 2, AO.

26 MacLeod to R.J. Richardson, 7 November 1972, commenting
on "Fort William Archaeological Project: Proposed Excavation

Program for 1973" by Joyce A. Kleinfelder, Lakehead University, file Fort William Archaeology 1972, RG 47-64 77-455, box 2, AO, emphasis in original.

There had been a native encampment and interpretation at Fort William from 1977. The changing role and status of Indigenous history at historic sites is discussed by scholars such as Robert Coutts and Laura Peers. Peers sees a significant improvement at Fort William from the mid-1990s and judges it unusually good for Indigenous engagement and representation. *Playing Ourselves: Interpreting Native Histories at Historic Reconstructions* (Lanham, MD: AltaMira Press, 2007), 15–16.

27 National Heritage, *Hinge of a Nation*, 79, 93, 82, and Appendix F.
28 Ibid., 94–6.
29 MacLeod, Supervisor of Research, and David Helliwell, Information Officer, to Bowes, Director, Historic Sites Branch, 6 February 1974, file F-NA-65, RG 47-64 77-455, box 1 [Fort William], AO; "A Disneyland of Northern Ontario?" *Globe and Mail*, 4 February 1974, 4; Stagg, personal correspondence, 3 May 2007.
30 Woolsey, "An Interpretative Framework," 6; Woolsey, Fort William Visitor Centre Planning Committee, to Bowes, 26 October 1973, file F-IN-77 Interpretative Programme 1973, RG 47-64 77-455, box 1, AO.
31 Woolsey to Dr. Fred Armstrong, University of Western Ontario, 26 September 1972, file F-NA-65, RG 47-64 77-455, box 1, AO. J.M.S. Careless, based at the University Toronto, was one of the most prominent Canadian historians in the twentieth century and involved in both the federal Historic Sites and Monuments Board and the provincial Archaeological and Historic Sites Board.
32 *Michipicoten Wilderness Area Action Plan 1972*, confidential, 16 March 1972, given to R.G. Bowes, and MacLeod to Bowes, 30 June 1971, file 4.7 Michipicoten Mission, RG 47-64 77-455, box 18, AO.
33 Woolsey to Bowes, Superintendent, Research and Planning Section, 18 April 1973, file 4.1 Temagami District, RG 47-64 77-455, box 18, AO, emphasis mine. In 1972, the Ministry of Industry and Tourism subsumed the old Department of Tourism and

Information, which had been responsible for initiating the Fort William project.

34 "Catering Services" and "Overnight Accommodations," Fort William Historical Park, accessed 29 January 2015, www.fwhp.ca. In *Superior Rendez-Vous*, Morrison indicates that Fort William moved from a program- to a market-oriented thinking in the 1990s (136–7), but I would suggest that this foundational strata was never dislodged in the first place.

35 I thank my colleagues in the department of history at Bucknell for their thoughts on this point.

36 Indeed, in 1973 the Branch applied to purchase adjacent properties along the river to further eliminate intrusive sights and sounds. Ministry of Natural Resources, "Application and Report to Management Board," 26 October 1973, file F-LA-55, RG 47-64 77-455, box 1, AO.

37 A. George Tracey, Geomorphologist, Park Planning Branch to [R.G. Bowes] Director, Historic Sites Branch, 15 October 1973, file F-IN-77, RG 47-64 77-455, box 1, AO.

38 This investment may not have diminished much. In 2010, the Hudson's Bay Company launched an advertising campaign themed "We Were Made for This" to coincide with the winter Olympics, presenting Canadians as the inheritors of a northern wilderness tradition born of the fur trade. See also Claire Campbell, "'It Was Canadian, Then, Typically Canadian': Revisiting Wilderness at Historic Sites," *British Journal of Canadian Studies* 21, no. 1 (2008): 5–34.

39 Ron Cockrell, *Grand Portage National Monument Administrative History* (National Park Service, 1983), 42. There had been a partial reconstruction at Grand Portage from the 1930s, and more systematic excavation and reconstruction through the 1960s and 1970s (slightly preceding that at Fort William). Carolyn Gilman and Alan Woolworth, *The Grand Portage Story* (St Paul: Minnesota Historical Society Press, 1992).

40 James Cleland Hamilton, *The prairie province; sketches of travel from Lake Ontario to Lake Winnipeg, and an account of the geographical position, climate, civil institutions, inhabitants,*

productions and resources of the Red Valley (Toronto: Belford
Bros., 1876), 5, 8; Patricia Jasen, *Wild Things: Nature, Culture,
and Tourism in Ontario, 1790–1914* (Toronto: University of
Toronto Press, 1995), 80, 100.

41 Duncan Campbell Scott, "The Height of Land," in *The Poems of
Duncan Campbell Scott* (Toronto: McClelland and Stewart, 1926),
45. Indeed, the watershed literally curls around Fort William to
Grand Portage.

42 Stephen Leacock, "I'll Stay in Canada," in *Funny Pieces: A Book
of Random Sketches* (New York: Dodd, Mead 1936), 291.

43 Ontario Department of Lands and Forests, "Mattawa: Wild River
Park," undated memo, file 4.2 Mattawa Wild River Park – General, RG 47-64 77-455, box 7, AO.

44 William Morton, *The Canadian Identity* (Madison: University of
Wisconsin Press, 1961), 5. Indeed, another name for those who
paddled between Montreal and Fort William was the "comers-and-
goers." These men, though, had lesser status than the *hivernants*
or over-winterers of the interior.

45 Douglas LePan, "Rough Sweet Land," in *Weathering It: Complete
Poems, 1948–87* (Toronto: McClelland and Stewart, 1987), 217–23.

46 Pierre Elliott Trudeau, "Exhaustion and Fulfillment: The Ascetic in
a Canoe," originally published in *Jeunesse Étudiante Catholique*
(1944), published in English in Borden Spears, ed., *Wilderness
Canada* (Toronto: Clarke, Irwin, 1970).

47 Pierre Berton, *Why We Act Like Canadians: A Personal Exploration of our National Character* (Toronto: McClelland and Stewart, 1982), 95–7, 108–9.

48 Franchère, *Narrative of a Voyage*, 338. In *A Preliminary Archaeological Investigation*, Dawson noted that the archaeological dig
often had to contend with the high water table and flooding.

49 William Van Horne to Hector Langevin, 5 March 1886, in *Thunder Bay District, 1821–1892: A Collection of Documents*, ed.
Elizabeth Arthur (Toronto: Champlain Society, 1973), 220.

50 Franchère, *Narrative of a Voyage*, 338–9.

51 Morrison, *Superior Rendez-Vous*, 58–9; Podruchny, *Making the
Voyageur World*, 36, 301. Food supply lay at the heart of the conflict between the Nor'westers and Selkirk; in 1814, the governor of

his Red River settlement, Miles Macdonnell, issued the "Pemmican Proclamation," which prohibited any fur traders not connected with the settlement from taking foodstuffs out of the area: an effectual blockade for NWC brigades. See Leslie Ritchie, "'Expectations of Grease & Provisions': The Circulation and Regulation of Fur Trade Foodstuffs," *Eighteenth-Century Life* 23, no. 2 (1999): 124–42. In the twenty-first century, the provisioning of northern communities has again become a point of concern, but for climactic rather than political reasons: with the winter ice roads forming later and later in the fall and melting faster in the springs, the supply routes to these communities are weakening.

52 George Mackenzie, "The Upper Lakes," in *Picturesque Spots of the North*, ed. George Grant (Chicago: Alexander Belford & Co., 1899), 90.

53 Martin E. Weaver, "Structural Conservation of the Buildings of Old Fort William Thunder Bay, Ontario," *Bulletin of the Association for Preservation Technology* 10, no. 3 (1978): 21–32.

54 Architectural historian Denis Mahon to Bowes, Director, Historic Sites, 6 March 1974, 5, file F-AU-65, RG 47-64 77-455, box 2, AO. The 1971 contract with National Heritage contained what came to be known as the "authenticity clause" [article 6(x) in the contract], which stated that if the Ministry was not satisfied with the historical accuracy of any part of the reconstruction, it could request that it be redone at no extra cost: "appearance of all aspects of the project shall be as historically accurate as possible, and in accordance with this intention, the most modern reconstruction and preservation techniques shall be used wherever possible ... If the Owner is of the opinion that any part of the erected work is not historically accurate (and the Project Manager cannot prove otherwise to the Owner's satisfaction) the Owner may, within six (6) months of the completion of construction of the particular phase, direct that such part of the work be redone to his satisfaction, at no additional cost to the Owner." The Historic Sites Branch wanted the evaluation carried out before the government assumed responsibility for the site.

55 J.W. Keenan, Executive Director, Parks Division, to A.J. Herridge, Assistant Deputy Minister, Resources and Recreation, 15 July 1974; Fort William Authenticity Committee minutes of 5 July 1974,

Thunder Bay; both file F-AU-65, RG 47-64 77-455, box 2, AO. Here, though, the committee could make only relatively superficial suggestions about such things as woodchips on park paths or a better approximation of eighteenth-century forest clearance. Kleinfelder to Piggott, undated; 5 January 1972 in "Fort William Authenticity Committee: Chronological Synopsis," 25 June 1974; and minutes of 5 July 1974, file F-AU-65, RG 47-64 77-455, box 2, AO. The committee's relative inability to suggest landscape revisions testifies to the difficulty of reconstructing an historic environment.

Critics recognized that reconstructed fur trade posts, which proliferated in the postwar period, could appear "too neat." One railed against historic sites (and their handlers) that "suffer from a congenital drive to make them look beautiful ... Forts really were not romantic. They did not have lawns fed with Vigoro stretching down to the river ... One wonders whether a Hudson's Bay factor, let alone an Indian or a coureur-de-bois would recognize it today." R.A.J. Phillips, "The Heritage Approach," in *The Canadian National Parks: Today and Tomorrow Conference II, Ten Years Later*, ed. J.G. Nelson (Waterloo, ON: Faculty of Environmental Studies, University of Waterloo, 1979), 687–97.

56 Piggott to Richardson, 20 August 1973, in "Fort William Authenticity Committee: Chronological Synopsis," 24 June 1974, file F-AU-65, RG 47-64, box 2, AO.

57 Parks Canada, *Rocky Mountain House National Historic Site Management Plan* (Ottawa: Parks Canada, 2007), 8, 14; Gary Forma, *Michipicoten Archaeology 1971: Investigations at the Hudson's Bay Company Fur Trade Post* (Place: Ontario Department of Lands and Forests, 1972), 4; Kevin Lunn, "York Factory National Historic Site of Canada: Planning the Future for a Place with a Momentous Past," *Manitoba History* 48 (2004): 23; Parks Canada, *York Factory National Historic Site Management Plan* (Ottawa: Her Majesty the Queen in Right of Canada, 2007), 5, 28; Dawson, *Preliminary Archaeological Investigation*, 73.

58 Jay Anderson, *Time Machines: The World of Living History* (Nashville, TN: American Association for State and Local History, 1984), 69. Anderson was a member of the American Association of Museums' Outdoor Museum Accreditation Committee. It is impor-

tant to acknowledge that Jean Morrison was hired as a staff historian in 1975, and after these first turbulent years the fort was able to establish a world-class library and archives of fur trade materials.

CHAPTER FOUR

1 To an environmental historian, the use of a biblical phrase to sanction settler dominion – literally, the Dominion of Canada – over nature is most telling. Bill C-529 was sponsored by Rick Laliberte, a Métis from northern Saskatchewan. Bill C-529, *An Act Respecting the Motto of Canada*, 3d sess., 37th Parliament, 2004, http://www.parl.gc.ca/HousePublications/Publication.aspx?Language=E&Mode=1&DocId=2334190.

2 Cited in Solomon Aremu, Richard Brundrige, Jeffery Lowe, and Ioannis Ziotas, *C.N.R. East Yards Redevelopment '84: A Showcase of Winnipeg's Past and Future* (Winnipeg: Institute for Urban Studies, University of Winnipeg, 1986), 90. Histories of the Forks can be found in Parks Canada, *The Forks National Historic Site Management Plan* (Ottawa: Her Majesty the Queen in Right of Canada, 2007); Manitoba Historic Resources Branch, *The Forks of the Red and Assiniboine Rivers: An Outline of Historical Significance* (Winnipeg: Manitoba Culture, Heritage and Recreation, n.d.); Parks Canada, *The Forks National Historic Site: Historic Themes* (Winnipeg: Department of Canadian Heritage, 1995); Randy R. Rostecki, "From Backwater to Park: The Forks in Relation to Downtown Winnipeg," in *The Forks and the Battle of Seven Oaks in Manitoba History*, ed. Robert Coutts and Richard Stuart (Winnipeg: Manitoba Historical Society, 1994). Parks Canada research reports include Robert Coutts, *The Forks of the Red and Assiniboine: A Thematic History, 1734–1850* compiled with Diane Payment, *Native Society and Economy in Transition at The Forks, 1850–1900* (Ottawa: Environment Canada, 1988); Gerry Berkowski, *The Forks Post 1870: Storyline* (Ottawa: Environment Canada, 1987); Rodger Guinn, *The Red-Assiniboine Junction: A Land Use and Structural History, 1770–1980* (Ottawa: Parks Canada, 1980).
    Despite the image of Winnipeg as a gateway to the grain-filled and golden prairie, several of these studies note that the river val-

leys were more of an approximation of the wooded parkland to the north. For a wonderful environmental history of the parkland ecoregion, see Merle Massie's *Forest Prairie Edge: Place History in Saskatchewan* (Winnipeg: University of Manitoba Press, 2014).

3 Splendid wide-ranging histories of the histories of the northwestern plains in this period can be found in Theodore Binnema, *Common and Contested Ground: A Human and Environmental History of the Northwestern Plains* (Norman: Oklahoma University Press, 2001) and George Colpitts, *Pemmican Empire: Food, Trade, and the Last Bison Hunts in the North American Plains, 1780–1882* (Cambridge: Cambridge University Press, 2014). Archaeology at the Forks led by the Canadian Parks Service suggests a three-thousand-year variable occupation, although with seasonal and temporary settlements suggesting a highly contested area. See Sid Kroker and Pam Goundry, *Archaic Occupations at The Forks* (Winnipeg: Forks Public Archaeological Association, 1994), which includes reports from the public archaeology projects in the early 1990s; and S. Biron Ebell, "The Red and Assiniboine Rivers in Southern Manitoba: A Late Prehistoric and Early Historic Buffer Zone?," *Manitoba Archaeological Quarterly* 12, no. 2 (1988): 3–26. David Meyer and Paul Thistle analyze certain locations along the Saskatchewan River as seasonal gathering places for Plains Cree in "Saskatchewan River Rendezvous Centers and Trading Posts: Continuity in a Cree Social Geography," *Ethnohistory* 42, no. 3 (1995): 403–45.

4 Lord Cultural Resources Planning and Management Inc., *Second Interim Report: The Forks Heritage Interpretive Plan* (Winnipeg: Lord Cultural Resources Planning and Management Inc., 1990) 2.5–2.6. Colpitts characterizes the Red River settlement as an "odd agricultural outpost of empire" (*Pemmican Empire*, 194) – a testament to the power and scale of the buffalo hunt until the mid-nineteenth century. The Forks North Portage Partnership has compiled a very helpful bibliography at http://www.theforks.com /about/history/heritage-research/bibliography.

5 Beginning in the 1970s, the Exchange District emerged as a leader in urban heritage renewal, preserving some of the best examples

of prewar "skyscrapers" and commercial architecture in Canada.
It too was designed to rejuvenate a flagging urban core. Parks
Canada, "Commemorating Winnipeg's Newest National Historic
Sites: The Exchange District and Union Bank Building," *Manitoba
History* 38 (1999–2000): 26–9; Matthew Komus, Jocelyne An-
deres, and Winnipeg Historical Buildings Committee, *Winnipeg's
Exchange District: A Heritage Guide* (Winnipeg: Exchange District
Business Improvement Zone in partnership with the City of Win-
nipeg Historical Buildings Committee, 2006).

6 Heritage Winnipeg, *Brief to the Planning Committee [City of Win-
nipeg] Regarding the Forks, March 29 1990*, Historic Resources –
The Forks Development (1990–1991), accession no. GR3487 N-10-
6-152, Archives of Manitoba (hereafter AM); Aremu et al., *C.N.R.
East Yards*, 1; Reid, Crowther & Partners, *A Market Analysis for
Metropolitan Winnipeg* (Winnipeg: Reid, Crowther & Partners,
1967). For example, the National Cartage Building (later the
Johnston Terminal) was the largest warehouse in Winnipeg by the
1930s, but sat empty from 1977 onwards.

7 Other designations at the Forks by the HSMBC included the Battle
of Seven Oaks (between Selkirk settlers and Métis and Nor'westers, 1920); Fort Douglas (1924); and Union Station (1976). Daniel
MacFarlane has a useful summary of the role of rivers in Canadian
historiography in *Negotiating a River: Canada, the US, and the
Creation of the St. Lawrence Seaway* (Vancouver: University of
British Columbia Press, 2014). Interestingly, Brian Woolsey, who
had been so instrumental at Fort William, wrote an early report on
interpretation possibilities at the Forks for Parks Canada. Not sur-
prisingly, he recommended that it should relate to the site to the
wider region and to other North West Company posts, including
Fort William. *An Interpretive Centre at the Forks: A Preliminary
Discussion of the Rationale, Requirements and Ramifications*
(Winnipeg: Parks Canada, 1975) and *Historical Resources of the
Red-Assiniboine: A Preliminary Analysis of their Interpretive and
Development Potential* (Winnipeg: Parks Canada, 1975).

8 On the development of heritage in the Prairie provinces in the
twentieth century, see James Opp, "Prairie Commemorations and

the Nation: The Golden Jubilees of Alberta and Saskatchewan,
1955," in *Canadas of the Mind: The Making and Unmaking of
Canadian Nationalisms in the Twentieth Century*, ed. Adam
Chapnick and Norman Hillmer (Montreal and Kingston: McGill-
Queen's University Press, 2007); Frances Kaye, *Hiding the
Audience: Viewing Arts and Arts Institutions on the Prairies* (Ed-
monton: University of Alberta Press, 2003); Mark Rasmussen,
"The Heritage Boom: The Evolution of Historical Resource Con-
servation in Alberta," *Prairie Forum* 15, no. 2 (1990): 235–62;
David Smith, "Celebrations and History on the Prairies," *Journal
of Canadian Studies* 17, no. 3 (1982): 47–57.

9 Pierre Trudeau, "Exhaustion and Fulfillment: The Ascetic in a
Canoe" (1944), published in English in *Wilderness Canada*, ed.
Borden Spears (Toronto: Clarke, Irwin, 1970), 5; ARC Management
Board, *Red River Corridor Master Development Plan* (September
1981); Aremu et al., *C.N.R. East Yards*, 41–2.

10 Garry Hilderman and Associates, *The Red & Assiniboine Rivers
Tourism & Recreation Study: phase 2* (Winnipeg: Hilderman and
Associates, 1975); ARC Branch, *The Red River Corridor: A Canada-
Manitoba ARC Proposal* (Ottawa: Planning Division, Parks
Canada, Department of Indian and Northern Affairs, 1976), 29.

11 ARC Management Board, *Red River Corridor Master Development
Plan* (September 1981), 13; Lombard North Group Ltd., *The
Forks Site Development Plan* (Winnipeg: Prepared for Environment
Canada, Parks Canada Prairie Region, 1986), 33; Aremu et al.,
C.N.R. East Yards, 46.

12 Lombard North Group Ltd., *The Forks Site Development Plan*, 10;
on viewscapes, see Lord Cultural Resources Planning and Management Inc., *Second Interim Report*, especially 5.5. Paul Downie
summarizes the fifty or so archaeological reports on the site, most
of which encountered industrial fill, in *The Forks National Historic
Site of Canada: Cultural Resource Inventory and Cumulative
Impacts Analysis* (Winnipeg: Report prepared for Manitoba Field
Unit, on file, Cultural Resource Services Unit, Western Canada
Service Centre, Parks Canada, 2002).

13 Andrew Hurley, "Narrating the Urban Waterfront: The Role
of Public History in Community Revitalization," *The Public*

Historian 28, no. 4 (2006): 21–2; IBI Group, "Site Development Plan Proposed National Historic Park at The Forks" (Winnipeg: prepared for Parks Canada, 1985), 3.

14 Susan Buggey, "The Halifax Waterfront Buildings: A Restoration Project" (Ottawa: National and Historic Sites Service, 1972); Halifax Planning Department, *Halifax Waterfront Development Area Plan* (Halifax: Halifax Planning Department, 1976); Canadian Register of Historic Places, "Halifax Waterfront Buildings National Historic Site of Canada," Statement of Significance, accessed 15 February 2017, http://www.historicplaces.ca/en/rep-reg/place-lieu.aspx?id=1614.

15 Development at the Forks stalled through the early 1980s in large part due to disagreements over who would buy the land from CN, and how much the government was willing to pay. Peter St John describes the political dealing that broke the logjam in "The Forks Today," in *Crossroads of the Continent: A History of the Forks of the Red & Assiniboine Rivers*, ed. Barbara Huck (Winnipeg: Heartland, 2003), 146–73. Aremu et al., *C.N.R. East Yards Redevelopment '84* summarizes a series of designs for the Forks that had been proposed in the 1970s; most featured mixed development (retail, public park, pedestrian access) but at least one imagined it as "a slice of midtown Manhattan" with office and apartment towers (31).

16 The major set of rail lines were left in place, forming the western border for the site. "Update: The Forks/CN East Yard August 1988," in Federal Government and National Organizations – East Yards/The Forks Park Planning (1987–1988), accession no. GR2516, location code K-7-2-15, AM.

17 Tourism Centre Development Committee, "Tourism Centre at the Forks: Project Summary," February 1989; Donna Dul, director, Historic Resources Branch, "Committee Submission to Treasury Board from Industry, Trade and Tourism," *Briefing Notes: Culture, Heritage & Recreation – Industry, Trade & Tourism – Project Development, the Forks* [21 October 1988], Federal Government and National Organizations – East Yards/The Forks Park Planning (1987–88), accession no. GR2516, location code K-7-2-15, AM; Aremu et al., *C.N.R. East Yards*, 66–71.

18 Christopher Dafoe, "Forking out at Historic Forks," *Winnipeg Free Press*, 10 September 1988; Val Werier, "History Unfolds at the Forks," *Winnipeg Free Press*, 25 January 1989.

19 During one presentation by the CEO of the Forks North Portage Partnership, a member of the audience stood up to say, "You must have misheard us, Mr. Diakiw. We were asking for more park, not more parking." Quoted in St John, "The Forks Today," 165–6.

20 Ian Chodikoff, "Pieces at Play," *Canadian Architect*, 1 March 2004, 14–18.

21 Illustrative of the "attraction" thinking is "Winnipeg Attractions – Program 3, Work Plan, Manitoba Place at the Forks," Federal Government and National Organizations – East Yards/The Forks Park Planning (1987–88), Deputy Minister of Culture, Heritage, and Tourism office files, accession no. GR2516, location code K-7-2-15, AM. In response, see Heritage Winnipeg, "Brief to the Planning Committee regarding the Forks, March 25, 1990," "Historic Resources – The Forks Development," Deputy Minister of Culture, Heritage, and Tourism office files, accession no. GR3487, location code N-10-6-152, AM.

22 Antoine Predock, Canadian Museum of Human Rights, design statement, 2014, http://www.predock.com/CMHR/CMHR.html.

23 Falk Environmental Inc., *Canadian Museum for Human Rights, Winnipeg, Manitoba: Environmental Assessment Report* (Winnipeg: Falk Environmental Inc., 2006). Site plans from twenty years before had, in fact, recommended against high-rise buildings "so as not to 'wall off' river vistas to non-residents ... Nor would one desire to see a permanent shadow cast over the riverfront promenade." Aremu et al., *C.N.R. East Yards*, 76.

24 On park facilities, see Parks Canada, Prairie Region, "Interpretative Plan Framework" (November 1985), 6; Appendix 1 in Lombard North Group Ltd., *The Forks Site Development Plan*. Regarding Fort Garry, see ARC Branch, *The Red River Corridor*, 26.

25 David Smyth, *The Fur Trade Posts at Rocky Mountain House*, manuscript report, no. 197 (Ottawa: Parks Canada, National Historic Parks and Sites Branch, 1976), 149; William C. Noble, *The Excavation and Historical Identification of Rocky Mountain House* (Ottawa: National Historic Sites Service, National and

Historic Parks Branch, Department of Indian Affairs and Northern Development, 1973); National Historic Sites Service, *Rocky Mountain House National Historic Park Provisional Development Plan* (Ottawa: Parks Canada, Department of Indian Affairs and Northern Development, 1969), 12; M.B. Payne and C.J. Taylor, "Western Canadian Fur Trade Sites and the Iconography of Public Memory," *Manitoba History* 46 (2003–04): 2–14.

26 For more on the archaeological digs, see Downie, *The Forks National Historic Site of Canada.*

27 Crocus Heritage Resource Planning, Hilderman, Witty, Crosby, Hanna, and Associates and Prairie Habitats, *Time and the River: A Conceptual Interpretive Plan for the Core Area Riverbanks of Winnipeg* (Winnipeg: Crocus Heritage Resource Planning, 1989), 2. I am indebted to the Manitoba Museum for providing me with a rare copy of this report. Crocus Planning Group was led by Graham A. MacDonald, who later authored several reports on western historic sites and landscapes, particularly in Alberta. Garry Hilderman and his architectural firm – now HTFC Planning and Design – has been involved with the Forks since the mid-1980s as project manager and designer for various elements, including the playgrounds, Oodena Celebration Circle, the Provencher Bridge, amphitheatre, and orientation node.

28 Parks Canada, *The Forks National Historic Site Management Plan*, 11. Various studies since the mid-1990s had observed that, "People who visit the site usually do so voluntarily in search of recreation and leisure. Hence, the Forks can be described as a leisure setting." Leo Pettipas and James Kacki, *The Forks Heritage Interpretative Plan* (Winnipeg: Forks Renewal Corporation, 1993), 73.

29 Meeting of Tom Symons, Chair of Historic Sites and Monuments Board, with Manitoba Heritage Committee, 24 October 1986, in Manitoba Heritage Council correspondence, schedule CH 0261, file Tom Symons material, box D 10 2 8, box 1, AM; Gerry Berkowski, Historic Resources Branch, "Preliminary Observations on Parks Canada's Major Visitor Reception Centres and Capital Restoration Projects, 1978/9–1988/9," 15 June 1987, file 400.11.B, "The Forks Complex," location code K-7-2-15, AM; Canadian Parks Service, Prairie & Northern Region, "The Forks: A Heritage Legacy,"

Winnipeg, 1990, 6, in "Historic Resources - The Forks Development," Deputy Minister of Culture, Heritage, and Tourism office files, accession no. GR3487, location code N-10-6-152, AM.

30 The Forks North Portage Partnership, *Focus on the Future: Concept and Financial Plan, 2001–2010* (Winnipeg: The Forks North Portage Partnership, 2001), 11.

31 Parks Canada, *The Forks National Historic Site Management Plan*, 11–13. The text is also featured on the Forks website.

32 Tracy Bowman, *Forks National Historic Site of Canada Outdoor Play and Learning Area Research Evaluation: Final Report: Revised* (Winnipeg?: Western and Northern Service Centre, Parks Canada, 2009); Monica Giesbrecht, principal, HTFC Planning & Design, personal correspondence, 16 June 2015.

33 The quotation is from C.G.D. Roberts's poem, "Tantramar Revisited," first published in *The Week*, 20 December 1883. On Fort Anne and Fort Edward, see Parks Canada, *Port-Royal, Fort Anne, Scots Fort and Fort Edward National Historic Sites of Canada Management Plan* (Hull, QC: Parks Canada, 2002), 50.

34 Rivers, and especially urban rivers, have been a wonderful focus for environmental history in recent years. Examples include Daniel MacFarlane, *Negotiating a River*; Jennifer Bonnell, *Reclaiming the Don: An Environmental History of Toronto's Don River Valley* (Toronto: University of Toronto Press, 2014); the essays in Stéphane Castonguay and Michèle Dagenais, eds., *Metropolitan Natures: Urban Environmental Histories of Montreal* (Pittsburgh: University of Pittsburgh Press, 2011); Martin Melosi, *Precious Commodity: Providing Water for America's Cities* (Pittsburgh: University of Pittsburgh Press, 2011); Christopher Armstrong, H.V. Nelles, and Matthew Evenden, *The River Returns: An Environmental History of the Bow* (Montreal and Kingston: McGill-Queen's University Press, 2009).

35 Crocus Heritage Resource Planning et al., *Time and the River*, 1. On the environmental themes, see pages 5–15; on site locations, pages 28–39.

36 Pettipas and Kacki, *The Forks Heritage Interpretative Plan*, 30, 43–54.

37 Parks Canada, *The Forks National Historic Site: Historic Themes*,
    14–15.
38 Downie, *The Forks National Historic Site*, and Falk Environmen-
    tal, *Canadian Museum for Human Rights*, both discuss the older
    riverine geographies. See also work by W.F. Rannie, such as "A
    Geomorphological Perspective on the Antiquity of The 'Forks,'"
    Manitoba Archaeological Journal 9, no. 1 (1999): 103–13.
39 Ron Williams, *Landscape Architecture in Canada* (Montreal and
    Kingston: McGill-Queen's University Press, 2014), 225–6.
40 Uwe Lübken, "Rivers and Risk in the City: The Urban Floodplain
    as a Contested Space," in *Urban Rivers: Remaking Rivers, Cities,
    and Space in Europe and North America*, ed. Stephane Castonguay
    and Matthew Evenden (Pittsburgh: University of Pittsburgh Press,
    2012), 130–44; Christof Mauch and Thomas Zeller, "Rivers in
    History and Historiography," in *Rivers in History: Perspectives
    on Waterways in Europe and North America*, ed. Christof Mauch
    and Thomas Zeller (Pittsburgh: University of Pittsburgh Press,
    2008), 3.
        On the Floodway see Robert Passfield, "'Duff's Ditch': The Ori-
    gins, Construction, and Impact of the Red River Floodway," *Mani-
    toba History* 42 (2001–02): 2–14; and Shannon Stunden Bower,
    Wet Prairie: People, Land and Water in Agricultural Manitoba
    (Vancouver: University of British Columbia Press, 2011), 142.
41 Rostecki, "From Backwater to Park"; Bonnell, *Reclaiming the
    Don*, especially chapters 2 to 4.

CHAPTER FIVE

1   George Colpitts, *Pemmican Empire: Food, Trade, and the Last
    Bison Hunts in the North American Plains, 1780–1882* (Cam-
    bridge: Cambridge University Press, 2014) 3; also Andrew C. Isen-
    berg, *The Destruction of the Bison: An Environmental History,
    1750–1920* (Cambridge: Cambridge University Press, 2000). Yet
    the memory of (or nostalgia for) the buffalo remained: American
    naturalist writers Frances Lee and Florence Page Jaques found the
    area west of Moose Jaw "a region of virgin prairie, and drove for
    many miles through desolate and uneven hills with hardly a sign of

human existence except our single line of road ... The landscape
needed buffalo to complete it. Instead we saw our first cowboy
who – O tempora, O shade of The Virginian! – was driving cattle
with a tractor." *Canadian Spring* (New York and London: Harper
& Brothers, 1947), 73, emphasis added.

The epigraph for this chapter is from Sheilagh S. Jameson, "Era
of the Big Ranches," *Alberta Historical Review* 18, no. 1 (1970): 1.

2 Alfred Crosby, *Ecological Imperialism: The Biological Expansion
of Europe, 900–1900* (Cambridge: Cambridge University Press,
1986). He notes that "the regions that today export more foodstuffs of European provenance – grains and meats – than any other
lands on earth had no wheat, barley, rye, cattle, pigs, sheep, or
goats whatsoever five hundred years ago" (7). For an example
explicitly focused on cattle, see John Ryan Fischer, "Cattle in
Hawai'i: Biological and Cultural Exchange," *Pacific Historical
Review* 76, no. 3 (2007): 347–72.

3 Language of calamity from Treaty Six, "between Her Majesty the
Queen and the Plain and Wood Cree Indians and other Tribes
of Indians at Fort Carlton, Fort Pitt and Battle River with Adhe-
sions," 1876. On First Nations in the ranching industry, see especially the work of Mary-Ellen Kelm, *A Wilder West: Rodeo in
Western Canada* (Vancouver: University of British Columbia Press,
2011); Morgan Baillargeon and Leslie Tepper, *Legends of Our
Times: Native Cowboy Life* (Vancouver: University of British Columbia Press in association with Canadian Museum of Civilization,
1998). John Thistle's *Resettling the Range: Animals, Ecologies,
and Human Communities in Early British Columbia* (Vancouver:
University of British Columbia Press, 2015) is a creative reading
of the biological and cultural displacement of Indigenous life in the
BC Interior, including through the mechanism of ranching.

4 Other cattle were acquired from Fort William, by then in Company
hands. Barry Kaye, "The Trade in Livestock between the Red River
Settlement and the American Frontier, 1812–1870," originally
published in *Prairie Forum* (1981), reprinted in Gregory P.
Marchildon, ed., *Business & Industry* (Regina: Canadian Plains
Research Centre, 2000), 3–25. Thomas Douglas, Earl of Selkirk,
fulsomely described his grant at the Forks of the Red River as

"immense open plains without wood, fine dry grass land, much of it capable of immediate cultivation and all well fitted for pasturage, particularly sheep." Samuel H. Wilcocke, Simon McGillivray, and Edward Ellice, Appendix no. 2, "Lord Selkirk's Advertisement and Prospectus of the New Colony," in *A narrative of occurrences in the Indian countries of North America, since the connexion of the Right Hon. the Earl of Selkirk with the Hudson's Bay Company, and his attempt to establish a colony on the Red River* (London: 1817), 7.

5 Henry Y. Hind, *North-West Territory: Reports of Progress together with a Preliminary and General Report, on the Assiniboine and Saskatchewan Exploring Expedition* (Toronto: John Lovell, 1859), 151–2.

6 David Breen, *The Canadian Prairie West and the Ranching Frontier 1874–1924* (Toronto: University of Toronto Press, 1974), 23.

7 J.W. Bengough, "Uncle Sam Kicked Out!" *Grinchuckle*, 23 September 1869. The central debate in Canadian ranching history has been genealogical rather than environmental: whether the Canadian ranching world was primarily the product of Anglophile interests from the East or if it was a (North) American practice. Older scholars (David Breen, Lewis Thomas) argued for the metropolitan framework; others (Terry Jordan-Bychkow, Max Foran) later emphasized a continental dynamic.

8 Cochrane to Minister of the Interior [John A. Macdonald], 10 February 1881, file 142709, vol. 1209, RG 15-D-V-1, Department of the Interior: Timber and Grazing Branch; Order in Council 23 December 1881, "Pasturage Lands," *Sessional Papers of the Dominion of Canada* 15, no. 30 (1882); Ron Williams, *Landscape Architecture in Canada* (Montreal and Kingston: McGill-Queen's University Press, 2014), 125–6. In 1882 alone, seventy-five leases were authorized, covering a total of more than four million acres. The estimated number of stock on the range rose from nine thousand to one hundred thousand head in the five years between 1881 and 1886. John C. Lehr, John Everitt, and Simon Evans, "The Making of the Prairie Landscape," *Prairie Forum* 33, no. 1 (2008): 1–38.

9 J.F. Fraser, *Canada as It Is* (London: Cassell and Company, 1905),

173. David Igler's study of ranching in California makes it clear that ranching was a business, indeed, a forerunner of twentieth-century agribusiness, in scale, labour, and environmental practices, in *Industrial Cowboys: Miller & Lux and the Transformation of the Far West, 1850–1920* (Berkley: University of California Press, 2001). Ranchers are often referred to in public memory as empire-builders (for example, by the US National Parks Service in their 1959 study, and the Friends of the Bar U on their website).

10 Simon Evans, *The Bar U Ranch and Canadian Ranching History* (Calgary: University of Calgary Press, 2004); Alan McCullough, "Not an Old Cowhand: Fred Stimson and the Bar U Ranch," in *Cowboys, Ranchers and the Cattle Business: Cross-Border Perspectives on Ranching History*, ed. Simon Evans, Sarah Carter, and Bill Yeo (Calgary: University of Calgary Press, 2000), 29–42.

11 Warren M. Elofson, *Frontier Cattle Ranching in the Land and Times of Charlie Russell* (Montreal and Kingston: McGill-Queen's University Press, 2004), 132–5. He restates this in later work, such as *So Far and Yet so Close: Frontier Cattle Ranching in Western Prairie Canada and the Northern Territory of Australia* (Calgary: University of Calgary Press, 2015), 23, 94. In Oregon, Nancy Langston observed that some ranchers in the 1990s referred to this "lackadaisical" style of ranching as "'Christopher Columbus ranching' – letting your cattle loose in the spring and then discovering them in the fall." The reference to Columbus reminds us again of the effect of large-scale ranching in colonialist occupation. Nancy Langston, *Where Land and Water Meet: A Western Landscape Transformed* (Seattle: University of Washington Press, 2003), 21.

12 Geoff Cunfer responds to earlier criticisms of farming exhaustion by suggesting agro-ecologies were relatively stable and sustainable, in *On the Great Plains: Agriculture and Environment* (College Station: Texas A&M University Press, 2005). For example, he argues that the range was not overstocked – given the ecological affinity between plains grasses and large grazers (bison, then cattle) – but rather undersupported with infrastructure, and only undermined by the expansion of cropland (53). In his estimation, it was not human intervention but variations in rainfall that most affected plains sustainability.

13 Max Foran suggests, for example, that Canadian ranchers over-grazed the plains to compensate for low beef prices on the American market, in "Crucial and Contentious: The American Market and the Development of the Western Canadian Beef Cattle Industry to 1948," *American Review of Canadian Studies* 32, no. 3 (2002): 451–76.

14 David Wrobel, *Promised Lands: Promotion, Memory, and the Creation of the American West* (Lawrence: University Press of Kansas, 2002), 11.

15 Theodore Roosevelt, *Ranch Life and the Hunting-Trail* (New York: The Century Co., 1888), 24.

16 Leroy Victor Kelly, *The Range Men: The Story of the Ranchers and Indians of Alberta* (Toronto: William Briggs, 1913), 11.

17 He added, "There's mighty little open range left, barring the mountains – it's all under wire now." Charles M. Russell, *Trails Plowed Under* (Garden City and New York: Doubleday & Co., 1927), 159–60, 163.

18 David Bright, *The Limits of Labour: Class Formation and the Labour Movement in Calgary, 1883–1929* (Vancouver: University of British Columbia Press, 1998), 1. Bright gives a good sense of the extent of urban life, manufacturing, and commerce in Calgary in the "ranching" era.

19 I thank Larry Pearson and colleagues at Old St Stephen's in Historic Resources Management for their insights on this. James Opp suggests that the commemorations marking fifty years of provincehood for Alberta and Saskatchewan noted a commercial imperative, both in seeing the value of history for tourism and celebrating "resource-based wealth," while invoking the guiding moral compass of the pioneer generation. "Prairie Commemorations and the Nation: The Golden Jubilees of Alberta and Saskatchewan, 1955," in *Canada of the Mind: The Making and Unmaking of Canadian Nationalisms in the Twentieth Century*, ed. N. Hillmer and A. Chapnick (Montreal and Kingston: McGill-Queen's University Press, 2007), 228.

20 William J. Byrne, "Finding the Funding and Other Provincial Responsibilities: The Alberta Experience," in *The Place of History: Commemorating Canada's Past*, ed. Thomas H.B. Symons (Ottawa:

Royal Society of Canada, 1997), 237–43. The province "actively discouraged" federal attempts to purchase the Cochrane Ranch, which arguably ranked more prominently in the history of Canadian ranching, but had far fewer *in situ* resources remaining by the mid-1970s, accordingly to Harry Tatro's 1974 study for Parks Canada. Instead, Alberta purchased much of the site in 1972 and 1976 (although it was not until 1973 that provincial legislation provided for historic site status; in other words, the province began to acquire the land before it could name it an historic site). The site was cleared in 1976 and 1977 using, in part, prison labour (!), leaving little above ground, and designated a provincial historic resource in 1979. Despite a 1987 proposal for historic development not unlike that happening at the Bar U further south – reconstruction of several buildings for year-round interpretation, with a particular emphasis on the ecological setting – only the ranch house was reconstructed as a multimillion dollar Western Heritage Centre in 1996, and later turned over to the Town of Cochrane as a conference centre. Harry A. Tatro, "A Survey of Some Historic Ranches in Southern Alberta / Done by Harry A. Tatro during December 1973 and January and February 1974, " Canadian Parks Service, for the Historic Sites and Monuments Board of Canada, Ottawa, 1974, 16–17; Loraine Fowlow, *A Year in the life of the Cochrane Ranche: A Design Concept for Development* (Calgary: University of Calgary. Faculty of Environmental Design, 1987); Mark Rasmussen, "The Heritage Boom: The Evolution of Historical Resource Conservation in Alberta," *Prairie Forum* 15, no. 2 (1990): 247; J. Corkan and W. Truch, *Report on an Old Ranch House Located on the Original Cochrane Ranche Company Property to the Government of Alberta* (Edmonton: Department of Highways & Transport, 1969); Ken Mather, *A History of the Cochrane Ranch Site, 1884–1977* (Edmonton: Alberta Historic Sites Service, 1978), held by the Historic Resources Management Branch; Roderick J. Heitzmann, "The Cochrane Ranche Historic Site: Archaeological Excavations, 1977," Archaeological Survey of Alberta occasional paper no. 16, Alberta Culture Historic Resources Division, 1980. William Naftel, a historian for Parks Canada, also noted that the Cochrane family retained mineral rights to the land even as they disposed of the ranch, and so after the discovery of oil at Leduc in 1947 the property again became valuable. Naftel, "The Cochrane Ranch," Canadian Historic Sites, Occasional papers in Archaeology and History, no. 16, National Historic Parks and Sites Branch, Indian and Northern Affairs, Ottawa, 1977, 143.

21 Environment Conservation Authority, Conservation of Historical and Archaeological Resources in Alberta, Proceedings of the Public Hearings, May–June 1972 (Edmonton, 1972), 31; Historic Sites Service, *Master Plan for the Protection and Development of Prehistoric and Historic Resources within Alberta* (ca. 1981), 333–4.

In fact, there were unsuccessful rounds of negotiation on two other sites (the Flying E and the 7U Brown) well over a decade before negotiations began on the Bar U purchase. Parks Canada, "Bar U Ranch National Historic Site, Alberta, Commemorative Integrity Statement," November 2000, Appendix 1. Given the tenor of the relationship between Ottawa and Edmonton in the 1970s, charged by the politics of energy, it is not surprising that acquisition of site – an often serendipitous, often laborious, bureaucratic process at the best of times – was derailed.

22 The Grant-Kohrs Ranch was first identified in a 1959 study by the US National Parks Service of potential sites, *The Cattlemen's Empire*. See also Douglas C. McChristian, *Ranchers to Rangers: An Administrative History of Grant-Kohrs Ranch National Historic Site* (n.p.: National Park Service, 1977); John Milner Associates, *Grant-Kohrs Ranch National Historic Site, Deer Lodge, Montana, Cultural Landscape Report Part 1: Landscape History, Conditions, and Analysis and Evaluation* (Charlottesville, VA: John Milner Associates, 2004); Shapins Associates, "Grant-Kohrs Ranch National Historic Site Landscape: Cultural Landscape Inventory" National Park Service Intermountain Regional Office, 2007.

23 Tatro, "Survey of Historic Ranches," 37, 39.

24 A.B. McCullough, *The Ranching Industry in Canada: Report on Evaluation of Potential Sites for Commemoration Prepared for the Historic Sites and Monuments Board of Canada* (n.p.: 1989); Parks Canada, *Bar U Ranch National Historic Site of Canada Management Plan* (Ottawa: Her Majesty the Queen in Right of Canada,

1995, rev. ed. 2005). That said, the endorsement of the Bar U
was a long time coming, and only after the board revoked earlier
recommendations including the Flying E (another George Lane
ranch), Glengarry 44 (also owned by Pat Burns), and 7U Brown
(founded by former Bar U employees). Parks Canada, "Commemorative Integrity Statement," Appendix 1, 27–9.

25 Friends of the Bar U Historic Ranch Association, "The Legacy,"
accessed 15 February 2017, http://www.friendsofthebaru.ca. The
story told at the site by interpreters remains firmly on the side of
"cattlemen's empire" in relaying the ambitious expansions of the
Lane/Burns eras.

26 Donald Worster calls the open range era an "unmitigated disaster,"
and indicative of the profoundly unsustainable (North) American
approach to western land management and land holding, in "Cowboy Ecology," in *Under Western Skies: Nature and History in
the American West* (New York: Oxford University Press, 1992),
34–52. See also the work by Warren Elofson, such as "Grasslands
Management in Southern Alberta: The Frontier Legacy," *Agricultural History* 86, no. 4 (2012): 143–68.

27 Evans, *The Bar U Ranch*, 299; Environment Canada, *Canada's
Green Plan for a Healthy Environment* (Ottawa: Environment
Canada, 1990), 89–90, emphasis mine; George Hoberg and
Kathryn Harrison, "It's Not Easy Being Green: The Politics of
Canada's Green Plan," *Canadian Public Policy/Analyse de poli-
tiques* 20, no. 2 (1994): 119–37; D.B. Dalal-Clayton, *Getting to
Grips with Green Plans National-Level Experience in Industrial
Countries* (London: Routledge, Taylor and Francis, 2013).

28 See, for example, Parks Canada, "Commemorative Integrity
Statement."

29 Parks Canada, *Bar U Ranch National Historic Site Management
Plan* (1995), 8–9; rev. ed. (2005), 32.

30 Matt Dyce, "'The Gateway to the Last Great West': Spatial Histories of the Athabasca Landing Trail," *Canadian Historical Review*
94, no. 2 (2013): 200–1. This is an interesting example of another
ecoregion of Alberta – the parkland – framed as a hinterland first
for resource harvest and then for heritage tourism.

31 I suspect this is one reason it scores highly for "visitor satisfaction"

(Parks Canada, Annex 1 in *The State of Canada's Natural and Historic Places*, 2011, http://www.pc.gc.ca/eng/docs/pc/rpts/ elnhc-scnhp/2011/annex/annex01.aspx). Ranching is not mentioned in Robert Passfield's study of "Industrial Heritage Commemoration in the Canadian Parks Service," *IA: Journal of the Society for Industrial Archeology*, Part 1, 16, no. 2 (1990): 15–39 and Part 2, 17, no. 1 (1991): 33–67. It is interesting, given Canada's ongoing commitment to fossil fuels, that the National Historic Sites System Plan's theme of "Developing Economies" stresses, in particular, how "The land's mineral wealth has shaped Canadian history."

32 Ronald Rees, *New and Naked Land: Making the Prairies Home* (Saskatoon: Western Producer Prairie Books, 1988), 146–9. Paul Starrs writes that "Ranching represents a valued closeness of people to nature that is all too seldom now achieved," in "An Inescapable Range, or the Ranch as Everywhere," in *Western Places, American Myths*, ed. Gary Hausladen (Reno: University of Nevada Press, 2003), 58.

33 Owen Wister, "The Evolution of the Cow-Puncher," *Harper's Monthly* (September 1895): "The knight and the cowboy are nothing but the same Saxon of different environments, the nobleman in London and the nobleman in Texas" (606); Ralph Connor, *The Sky Pilot: A Tale of the Foothills* (Chicago and Toronto: Fleming H. Revell Co., 1899), 26. Of course, this blended well with the argument that the (anglo-)Canadian cowboy was distinct thanks to his British heritage, and inherently something of a gentleman, as per the *Calgary Herald* (12 November 1884, ten years before Wister's article): "Calgary is a western town, but not ... in the ancient sense of the word. It is peopled by native Canadians and Englishmen, who own religion and respect law. The rough and festive cowboy of Texas and Oregon has no counterpart here ... The genuine Alberta cowboy is a gentleman." This was reinforced when the Prince of Wales (later Edward VIII) visited the Bar U as part of his tour of Canada in 1919. Taken with the romance of the old-time ranching frontier, he purchased a ranch near High River – named "EP" for "Edward Prince" – which he owned until 1962. Alberta recently designated the "EP Ranch" a provincial historic

site. Simon M. Evans, *Prince Charming Goes West: The Story of the E.P. Ranch* (Calgary: University of Calgary Press, 1993); Alberta Community Development, Heritage Resource Management Branch, file des. 396; Canadian Register of Historic Places, "E.P. Ranch," accessed 1 March 2017, http://historicplaces.ca/en/rep-reg/place-lieu.aspx?id=1171.

34 See Paolo Pietropaolo, "The Park Paradox: Balancing Ecological Preservation and Human Use in the South Okanagan Valley" (master's thesis, University of British Columbia, 2016).

35 Ted Grant and Andy Russell, *Men of the Saddle: Working Cowboys of Canada* (Toronto: Van Nostrand Reinhold Ltd, 1978), 93. Most recently, this conflict, and these characterizations, reignited in Oregon in 2015, ostensibly over the Malheur Wildlife Refuge. In "The War for the West Rages On," *New York Times*, 29 January 2016, one of the protestors explains, "From the moment their ancestors' horses took a sip of water or ate the grass, 'a beneficial use of a renewable resource' was created." There are numerous case studies that illustrate the relationships (tensions and collaborations) between different constituencies in ranchland environments – ranchers, Indigenous peoples, government officials, etc. – over contested resources. Some examples include Adam M. Sowards, "Reclamation, Ranching, and Reservation: Environmental, Cultural, and Governmental Rivalries in Transitional Arizona," *Journal of the Southwest* 40, no. 3 (1998): 333–61; Zoltan Grossman, "Cowboy and Indian Alliances in the Northern Plains," *Agricultural History* 77, no. 2 (2003): 355–89; Joanna Reid, "The Grasslands Debates: Conservationists, Ranchers, First Nations, and the Landscape of the Middle Fraser," BC *Studies* 160 (2008): 93–118; Philip Garone, "Rethinking Reclamation: How an Alliance of Duck Hunters and Cattle Ranchers Brought Wetland Conservation to California's Central Valley Project," in *Natural Protest: Essays on the History of American Environmentalism*, ed. Jeff Crane and Michael Egan (New York: Routledge, 2009), 137–62.

36 Parks Canada, "Commemorative Integrity Statement," 23; also Evans, *The Bar U Ranch*, 304.

37 In "Grasslands Management," Elofson outlines some of more sustainable grazing practices recently instituted on the Walrond ranch

lands although he concludes that in many places the land is still exhausted from century-old overuse. For another, American, voice of rancher-as-steward, see an interview with Sid Goodloe, who ranches in New Mexico: "Ranching and the Practice of Watershed Conservation," in *Thinking Like a Watershed: Voices from the West*, ed. Celestia Loeffler and Jack Loeffler (Albuquerque: University of New Mexico Press 2012), 193–214.

In 2013, the Walrond was also the subject of a massive easement to the Nature Conservancy of Canada, as was the Chinook Ranch, south of the Bar U, in 2011. "Waldron Easement Conservation Easement on More Than 12,000 Hectares Largest in Canadian history," 29 September 2014, CBC *News*, http://www.cbc.ca/news/canada/calgary/alberta-ranchers-to-conserve-huge-tract-of-native-grassland-1.2781482; "Gift Preserves Alberta Land," *Calgary Herald*, 25 June 2011; Dave Ebner, "Ian Tyson to Lead Development Protest," *Globe and Mail*, 26 June 2006.

In 2002, the province of Alberta introduced a category of designation for "heritage rangeland" – alongside wilderness areas and ecological reserves – "in order to ensure its preservation and protection using grazing to maintain the grassland ecology." Alberta, Wilderness Areas, Ecological Reserves, Natural Areas and Heritage Rangelands Act, section 4.1, W-9 RSA 2000. This permits long-term leases and "traditional" ranching practices, and prohibits new mining permits, but does not extinguish any existing mineral claims and well sites. Such rangelands have been established to the north and south of the Bar U, including the OH Ranch Heritage Rangeland, privately run by the Calgary Stampede, and the Pekisko Heritage Rangelands, extending south from the Highwood River. At 34,356 hectares it represents about 85 percent the size of a large-lease ranch holding.

38 Donald Worster, "Two Faces West: The Development Myth in Canada and the United States," in *Terra Pacifica: People and Place in the Northwest States and Western Canada*, ed. Paul Hirt (Pullman: Washington State University Press, 1998), 76.

39 Medicine Hat Board of Trade, *Medicine Hat, Alberta: Facts Concerning the City and Surrounding Country* (Medicine Hat: Medicine Hat Board of Trade, 1918). Rudyard Kipling bequeathed

the city with the best tag line in the country when he described it
as having "all hell for a basement." "Wonder City of Canada,"
Calgary Herald, 4 October 1907, cited in David Jones, *Empire of
Dust: Settling and Abandoning the Prairie Dry Belt* (Edmonton:
University of Alberta Press, 1987), 29.

40 The first oil drilled in western Canada was actually within Water-
ton Lakes National Park. On Turner Valley, see the community
history *In the Light of the Flares: History of the Turner Valley Oil-
fields* (Turner Valley: Sheep River Historical Society, 1979); David
Finch, *Hell's Half Acre: Early Days in the Great Alberta Oil Patch*
(Surrey, BC: Heritage House, 1995). On the historic site, see Monenco Consultants Ltd., *Western Decalta Turner Valley Gas Plant
Site Characterization Report, 1987–1988* (Calgary: Monenco
Consultants Ltd., 1988); Alberta Community Development, *Development Plan: Turner Valley Gas Plant* (Edmonton: Alberta Community Development with Parks Canada, 2000). A disturbingly
long list of environmental assessments can be found via the provincial ministry of Culture and Tourism's Turner Valley Gas Plant
website at http://history.alberta.ca/turnervalley/reclamation/studies/
studies.aspx.

Not coincidentally, the Grant-Kohrs Ranch National Historic
Site in Montana is in a similar situation. The US Environmental
Protection Agency listed the Upper Clark Fork River as a Super-
fund site in 1984, citing a century of hazardous substances and
heavy metals from mining, milling, and smelting. The Grant-Kohrs
Ranch lies within the designated Superfund/National Priorities List
site boundaries and has been undergoing restoration since 2008.
These two examples remind us that "the frontier" was very much a
site of competing and coexisting land uses. The historical emphasis
has been on the encroachment of settler homesteads on ranch
lands, and the turning of the sod; but the relationship between
ranch lands and mining and fossil fuel industries – especially carrying forward to the present day – deserves more attention.

It is worth adding – to demonstrate this tangle of concurrence
and proximity – first, an adjacent community history of the same
era as that of Sheep River insisted "Stock-raising ... will always
continue to be a very important, if not the most important,

industry in a country so well adapted for it." Millarville, Kew, Priddis, and Bragg Creek Historical Society, *Our Foothills* (Calgary: Millarville, Kew, Priddis, and Bragg Creek Historical Society, 1975), n.p. Meanwhile, the province opened an Oil Sands Discovery Centre at Fort McMurray in 1985.

41 Wrobel, *Promised Lands*, 181–2; Ed Struzik, "Don't Tell These Ranchers Climate Change Isn't Real," *The Tyee*, 17 November 2015, http://thetyee.ca/News/2015/11/17/Alberta-Cattle-Ranges-Drought/. I am grateful to Margaret Herriman for sharing her story with me. An admittedly bleak survey of the ecological vulnerabilities of rangeland is Howard Gordon Wilshire, Jane E. Nielson, and Richard W. Hazlett, "Raiding the Range," *The American West at Risk: Science, Myths, and Politics of Land Abuse and Recovery* (Oxford: Oxford University Press, 2008), 77–99.

42 Douglas LePan, "Rough Sweet Land," in *Weathering It: Complete Poems 1948–87* (Toronto: McClelland and Stewart, 1987), 217.

CONCLUSION

1 A second irony is that as I write this I am living in central Pennsylvania. Since I am not going to become an American historian, I can't fall back on my usual practice of adopting regional historic sites to study. And the effect is palpable: I don't feel nearly as home here.

2 The framework of first and second nature is used by William Cronon, *Nature's Metropolis: Chicago and the Great West* (New York: W.W. Norton and Co., 1991), and Richard Judd, *Second Nature: An Environmental History of New England* (Amherst: University of Massachusetts Press, 2014).

3 John Wadland, "Loons and Landscapes: The Place of Environmental Heritage," in *The Place of History: Commemorating Canada's Past*, ed. Thomas Symons (Ottawa: Royal Society of Canada, 1997), 53. I am also inspired by Wadland's characterization of Canada as an unfinished project, in "Voices in Search of a Conversation: An Unfinished Project," *Journal of Canadian Studies* 35, no. 1 (2000): 52–75.

4 Here I would like to apologize to the editors of *Time and a Place: An Environmental History of Prince Edward Island* (Montreal and

Kingston: McGill-Queen's University Press; Charlottetown: Island
Studies Press, 2016), because they make a lovely argument for
studying islands as bounded, but connected, spaces.

5 I owe the expression "Hope and Ruin" to Nova Scotia band The
Trews (from their album of the same name, 2011).

6 This is found in Bengt Schonback and Nicolas Dykes, "L'Anse aux
Meadows National Historic Park: Interpretive Plan and Exhibit
Storyline," first draft, 22 April 1974, Ottawa. Commonalities be-
tween settler cultures is ubiquitous in H.J. Porter and Associates,
"Preliminary Development Concept: L'Anse aux Meadows
National Historic Park" (Halifax: Parks Canada, Indian and
Northern Affairs, 1975).

7 C.J. Taylor, *Negotiating the Past: The Making of Canada's National Historic Parks and Sites* (Montreal and Kingston: McGill-
Queen's University Press, 1990); Forum: Louisbourg Researchers
Recall Their Roles in the Reconstruction of Louisbourg, 1961–
2013, *The Nashwaak Review* 30–31, no. 1 (2013). This may extend to national parks, as well, even with the centennial of Parks
Canada in 2011. In 2008, Alan MacEachern suggested that parks
history was more marginalized than it had been at the original
"Parks for Tomorrow" conference forty years before, and there is
still more research on the national parks than on national historic
sites. Alan MacEachern, "Writing the History of Canadian Parks:
Past, Present, and Future" (paper presented at the Parks for Tomorrow 40th Anniversary Conference, Calgary, May 8–11, 2008).

Regarding the state of Parks Canada today, I read with interest
Cathy Stanton's observations about the institutional culture of the
American National Park Service, and its effect on both staff morale
and public opinion. As she writes, "more collegial critiques of the
Park Service ... have tended to stop short of asking the really hard
questions about how institutional culture may be getting in the
way of the good work that so many historians and others within
the agency truly want to do." "Does the National Park Service
Have a Culture Problem?," 19 July 2016, National Council of
Public History, http://ncph.org/history-at-work/does-the-national-
park-service-have-a-culture-problem/.

8 See Taylor, *Negotiating the Past*. Parks Canada, Indian and Northern Affairs, *National Historic Sites Policy* (Ottawa: Information
   Canada, 1972) – issued at the high point of historic park development – explains, "For the purposes of definition, a national historic
   park shall generally be considered to be an area with or without
   structures of major historic significance suitable in size for development as a park with effective interpretive displays" (3). The term
   national historic park was used as early as 1917 with the acquisition of Fort Howe and Fort Anne, but codified in the 1930
   National Parks Act (see the Introduction to this volume). The term
   seems to have faded out of use by Parks Canada: it is not referenced in the 1997 National Historic Sites *Systems Plan*, the 1998
   Parks Canada Agency Act, or the 2000 National Parks Act. But
   the 1982 National Historic Park Regulations still seem to be on the
   books (and which reference the 1930 Act), and more intriguing
   still, the term reappears in Parks Canada's *2014–15 Report on
   Plans and Priorities*, which announces a "New Guiding Narrative
   for History": "In 2014, Parks Canada will create a new framework
   for history at Parks Canada's national historic sites. This framework will be produced in consultation with federal partners.
   The second phase of this initiative will include the identification
   of national historic parks based on chapters of the historical
   framework." See http://www.pc.gc.ca/eng/docs/pc/plans/rpp/rpp
   2014-15/index.aspx. We await this narrative, and these parks.

Index